NUREG-1423
Volume 15

A Compilation of
Reports of
The Advisory
Committee on
Nuclear Waste

September 2004 -June 2006

U. S. Nuclear Regulatory
Commission

October 2006

ABSTRACT

This compilation contains 30 reports issued by the Advisory Committee on Nuclear Waste (ACNW) during the sixteenth and seventeenth years of its operation. The reports were submitted to the Chairman and the Executive Director for Operations of the U. S. Nuclear Regulatory Commission (NRC). Reports prepared by the Committee have been made available to the public through the NRC Public Document Room, the U. S. Library of Congress, and the Internet at http://www.nrc.gov/reading-rm/doc-collections.

PREFACE

The enclosed reports are the recommendations and comments of the U. S. Nuclear Regulatory Commission's Advisory Committee on Nuclear Waste during the period between September 1, 2004 and June 30, 2006. Previously issued Volumes 1 through 14 of NUREG-1423 contain the Committee's recommendations and comments from July 1, 1988 through August 31, 2004.

ACNW MEMBERSHIP (JULY 1, 2004 - JUNE 30, 2006)

CHAIRMAN: Dr. B. John Garrick, Consultant
St. George, UT
(Term ended: 10/04)

VICE CHAIRMAN: Dr. Michael T. Ryan, Consultant and Faculty Member
Charleston Southern University
Charleston, SC
(Chairman: 10/04 - present)

MEMBERS: Dr. James H. Clarke
Professor, Practice of Civil and Environmental Engineering
   and Director of Graduate Studies for Environmental Engineering
Vanderbilt University

Mr. Allen G. Croff (Retired)
Oak Ridge National Laboratory (ORNL)
(Vice Chairman: 10/04 - present)

Dr. William J. Hinze, Professor Emeritus
Purdue University

Dr. George M. Hornberger
Associate Dean for the Sciences
University of Virginia
Charlottesville, VA
(Term ended: 10/04)

Ruth Weiner, Consultant
Sandia National Laboratories
Albuquerque, NM

EXECUTIVE
DIRECTOR: Dr. John T. Larkins
Advisory Committee on Nuclear Waste
U. S. Nuclear Regulatory Commission

TABLE OF CONTENTS

TABLE OF CONTENTS (CONT'D)

TABLE OF CONTENTS (CONT'D)

November 2, 2004

MEMORANDUM TO: Luis A. Reyes
Executive Director for Operations

FROM: John T. Larkins, Executive Director
Advisory Committee on Nuclear Waste

SUBJECT: DRAFT FINAL REGULATORY GUIDE DG-1085, "STANDARD FORMAT AND CONTENT OF DECOMMISSIONING COST ESTIMATES FOR NUCLEAR POWER REACTORS," AND NUREG-1713, "STANDARD REVIEW PLAN FOR DECOMMISSIONING COST ESTIMATES FOR NUCLEAR POWER REACTORS"

During the 154th meeting of the Advisory Committee on Nuclear Waste (ACNW), October 19-21, 2004, the Committee considered for review draft final Regulatory Guide DG-1085, "Standard Format and Content of Decommissioning Cost Estimates for Nuclear Power Reactors," and draft final NUREG-1713, "Standard Review Plan for Decommissioning Cost Estimates for Nuclear Power Reactors." As noted in my October 12, 2004, memorandum to you, the Advisory Committee on Reactor Safeguards (ACRS) declined to review these documents and forwarded them to the ACNW.

At its 154th meeting, the ACNW similarly decided not to review these documents.

Reference:

Memorandum dated September 24, 2004, from Catherine Haney, Program Director, NRR, to John T. Larkins, Executive Director, ACRS, Subject: Publication of Regulatory Guide DG-1085, "Standard Format and Content of Decommissioning Cost Estimates for Nuclear Power Reactors," and NUREG-1713, "Standard Review Plan for Decommissioning Cost Estimates for Nuclear Power Reactors".

cc: A. Vietti-Cook, SECY
W. Dean, OEDO
R. Tadesse, OEDO
J. Dyer, NRR
C. Haney, NRR
M. G. Crutchley, NRR
C. Pittiglio, NRR
M. R. Snodderly, ACRS

UNITED STATES
NUCLEAR REGULATORY COMMISSION
ADVISORY COMMITTEE ON NUCLEAR WASTE
WASHINGTON, D.C. 20555-0001

November 3, 2004

The Honorable Nils J. Diaz
Chairman
U.S. Nuclear Regulatory Commission
Washington, D.C. 20555-0001

SUBJECT: WORKING GROUP ON THE EVALUATION OF IGNEOUS ACTIVITY AND ITS CONSEQUENCES FOR A GEOLOGIC REPOSITORY AT YUCCA MOUNTAIN, NEVADA

Dear Chairman Diaz:

During its 153rd meeting on September 22-23, 2003, the Advisory Committee on Nuclear Waste (ACNW) had a working group meeting (WGM) on the evaluation of igneous activity and its consequences for the potential Yucca Mountain high-level waste repository. The WGM included panel discussions by eight renowned scientists from academia, research institutions, and private enterprise in the fields of volcanism, risk assessment, and health physics[1]. Presentations to the Committee were made by the Nuclear Regulatory Commission (NRC) staff, staff from the Center for Nuclear Waste Regulatory Analyses (CNWRA), the Electric Power Research Institute (EPRI), ACNW staff, LANL, SNL, ORNL, the University of Utah, and ABS Consulting, Inc. Stakeholders and members of the public were given opportunities to comment on the discussions. Representatives from DOE's Yucca Mountain Project Office and the State of Nevada were invited to give presentations but declined.

The purposes of the WGM were to (1) increase the ACNW's technical knowledge of staff plans to evaluate the likelihood and consequences of disruptive igneous events at the proposed Yucca Mountain repository; (2) better understand NRC staff expectations regarding the DOE's consequence analyses; (3) identify aspects of those analyses that may need further study; and (4) complement previous working group meetings on performance assessments of Yucca Mountain. In addition, there were discussions regarding (1) the technical bases (measurements, analyses, and interpretations) necessary to conduct dose assessments, (2) the role of risk insights in the development of technical bases, and (3) the impact of outstanding technical issues on the resolution of agreements. The expert panel offered a number of suggestions and observations regarding the assessments and evaluations that will support the

---

[1]Drs. Bruce Crowe (the Los Alamos National Laboratory – LANL), William Hinze (Purdue University), Bruce Marsh (Johns Hopkins University), William Melson (Smithsonian Institution), Robert Budnitz [Lawrence Livermore National Laboratory, on detail to the U.S. Department of Energy (DOE)], Fred Harper (the Sandia National Laboratories – SNL), Lynn Anspaugh (University of Utah), Keith Eckerman (the Oak Ridge National Laboratory – ORNL), and ABS Consulting, Inc. (Irvine, California).

volcanism-related dose calculations. The calculations must be included in a DOE license application to meet the requirements of 10 CFR Part 63.

The WGM covered three areas of interest: (1) probability that future basaltic dikes will intersect a potential repository; (2) the manner in which a volcanic event intersects a waste disposal drift and mobilizes radioactive material from waste packages; and (3) the dosimetric consequences of subsequent dispersal of radioactive material.

To prepare for the WGM, ACNW staff and consultants attended an Appendix 7 meeting on September 21, 2004, between the NRC and DOE, where there was a presentation on the preliminary results of a recent aeromagnetic survey in the region, designed to detect possible additional buried basaltic features. Additionally, staff learned that DOE is reconvening an expert panel on volcanic hazards to examine new data that have become available since the previous 1996 expert elicitation.

Based on the information presented at the WGM, the Committee has concluded that it was not clear or transparent how the staff's work on igneous activity is risk informed. The Committee makes the following recommendations as a result of this WGM:

1. Instead of using a fixed value of 10^{-7} per year in performance assessments to represent the dike intersection frequency, it would be better to use an appropriate range, such as 10^{-8} to 10^{-7} per year, as suggested by Dr. Crowe, one of the WGM panelists. A similar range was derived in a recent ACNW staff paper (Coleman et al., 2004). This range is consistent with the Committee's previous conclusion in 2002. "The range of estimated probabilities, $\sim 10^{-9}$ to $\sim 10^{-7}$ per year, of an igneous intrusion into the repository used by DOE in its performance assessment is reasonable." Such a range is consistent with the volcanic history of the Yucca Mountain region.

2. The staff should give high priority to examining the realism in models for evaluating the potential interaction of magma with repository drifts and waste packages. The staff assumes that all of the radioactive material in a waste package becomes available after interaction with intrusive magma. The Committee heard an alternative view from EPRI. EPRI scientists presented an analysis of a postulated magma intrusion scenario, and contended that there is a "reasonable expectation" that no waste packages will fail during a postulated intrusive igneous event. The Committee believes that additional evaluation of waste package/magma interactions would improve the risk insights regarding the quantities of radioactive materials that could be mobilized. Recommendations provided by both EPRI (2004) and the DOE-sponsored Igneous Consequences Peer Review (ICPR) Group can offer insights on how to improve this modeling.

3. Based on the presentations, the Committee believes the staff should reassess the apparent conservatisms in the consequence and dose estimates from airborne transport of contaminated volcanic ash. Examples include wind direction, mass loading, and other parameters used in calculating dose to the reasonably maximally exposed individual (RMEI). A more transparent calculation would show how these assessments are risk informed.

4

Probability that future volcanism will intersect a potential repository

The most recent system-level performance assessment by the NRC staff (Mohanty et al., 2004) used a constant value of 10^{-7} per year for igneous intrusion rather than a range of probabilities. A new analysis of probability has been performed by the ACNW staff and its consultant and NRC's Office of Nuclear Research staff. This work (Coleman et al., 2004), which has been accepted for publication in the journal *Geophysical Research Letters*, suggests that an appropriate range for the likelihood of igneous intrusion into the repository is 10^{-8} to 10^{-7} per year. This range is identical to that reported in a paper by the NRC and CNWRA staff (Connor et al., 2000). These are two examples of a number of published evaluations of the likelihood of igneous events in the Yucca Mountain region. The Committee believes that a thorough, documented review of these and related evaluations will be useful in making staff analyses more transparent.

Volcanic event intersects a waste disposal drift and mobilizes radionuclides from waste packages

The NRC staff currently assumes in its modeling that the entire radioactive content of a waste package intercepted by intruding magma is available for airborne transport during a volcanic eruption. Representatives from EPRI discussed the impacts of potential igneous activity on waste packages. Their simulations suggested that under assumed conditions the waste packages would not be breached. Erosive effects of flowing magma were reported to be unlikely and waste packages did not fail from simulated overpressure effects or creep failure. EPRI representatives concluded there is "reasonable expectation" that no waste packages will fail during a postulated igneous event. EPRI only considered a scenario where the waste package had not been breached prior to magma intrusion into the repository. This may be reasonable, based on NRC staff analyses that show mean dose arising from extrusive igneous activity is much greater if the intrusion occurs in the first 500 years after postulated waste emplacement (Mohanty et al., 2004). In addition, the Committee heard presentations regarding the water content of magma which is important to its physical properties such as viscosity and explosivity (Nicholis and Rutherford, 2004). In November 2003, at its 147th meeting, the Committee was briefed on the DOE-sponsored ICPR Group recommendations (Detournay et al., 2003a and b). The ICPR was tasked to critically review the technical bases used by DOE to analyze the consequences of igneous events that might impact a repository, and to make recommendations on additional tasks that would significantly strengthen that program. Both EPRI's (2004) and the ICPR Group's recommendations offer insights on how to improve the consequence modeling. The Committee believes that it would be beneficial for the staff to consider these works in further evaluations of igneous intrusion scenarios.

Estimation of potential doses from igneous activity

The Committee heard presentations from NRC staff, the CNWRA, and other experts on the behavior of aerosols generated during explosive events involving metals and ceramics, resuspension modeling, internal dosimetry modeling and an independent comprehensive assessment of the consequence scenario. The Committee concluded from these presentations that the staff's assumptions and consequence modeling of an igneous event could be overly conservative in several ways:

1. It is unclear what fraction of the radioactive material could be involved in an eruption to which the RMEI is ultimately exposed. Further, an analysis of the range of values associated with release, transport, and exposure of radioactive material would improve the risk insights. For example, particle sizes up to 100 microns are included in the dose assessment, although 10 microns is typically considered the upper limit of the respirable range. The Committee heard during an expert presentation that in explosions designed to disperse metals and ceramics, typically less than 10 percent of the mass of particulate matter is smaller than 10 microns in diameter. The remainder of the particulate matter is larger and settles out quickly.

2. The current staff analysis assumes the wind blows towards the RMEI at all times. Always placing the receptor directly downwind artificially and incorrectly increases the estimated dose. The staff reported that this conservatism in transport and exposure modeling was being addressed, though results were not available. The re-analysis will consider a distribution of wind directions based on weather data from the Yucca Mountain region.

3. Assumed dust loadings are quite high and resuspension is modeled to continue for years. An expert panel member reported that resuspension is a phenomenon that is generally important for days after a release, rather than years. This conclusion was based on data from work at the Nevada Test Site during above-ground nuclear weapons testing.

These are examples of apparent conservatisms that result from fixed value assumptions. It is difficult for the Committee to see this as a realistic assessment. A systematic evaluation of ranges of parameters may provide more transparent risk insights in the ultimate calculation of dose to the RMEI.

Sincerely,

Michael T. Ryan
Chairman

Cited References

Coleman, N., B. Marsh, and L. Abramson, "Testing Claims About Volcanic Disruption of a Potential Geologic Repository at Yucca Mountain, Nevada," accepted for publication by *Geophysical Research Letters*, October 2004.

Connor, C.B., J.A. Stamatakos, D.A. Ferrill, B.E. Hill, G.I. Ofoegbu, F.M. Conway, B. Sagar, and J. Trapp, "Geologic Factors Controlling Patterns of Small-Volume Basaltic Volcanism: Application to a Volcanic Hazards Assessment at Yucca Mountain, Nevada," *Journal of Geophysical Research*, 105(B1): 417–432 [January 2000].

Detournay, E., L.G. Mastin, J.R.A. Pearson, A.M. Rubin, and F.J. Spera, "Final Report of the Igneous Consequences Peer Review Panel," Final Report of the Igneous Consequences Peer Review Panel," Las Vegas, Bechtel-SAIC Co., LLC., April 2003a. [Prepared for DOE.]

Detournay, E., L.G. Mastin, J.R.A. Pearson, A.M. Rubin, and F.J. Spera,, "Appendices to the Final Report of the Igneous Consequences Peer Review Panel," Las Vegas, Bechtel-SAIC Co., LLC., February 2003b. [Prepared for DOE.]

EPRI (Electric Power Research Institute), "Potential Igneous Processes Relevant to the Yucca Mountain Repository: Extrusive-Release Scenario. Analysis and Implications," Palo Alto, Final Report 1008169, June 2004.

Mohanty, S., et al., "System-Level Performance Assessment of the Proposed Repository at Yucca Mountain Using the TPA Version 4.1 Code," San Antonio, Center for Nuclear Waste Regulatory Analyses, CNWRA 2002-05 (Rev. 2), March 2004. [Prepared for the NRC.]

Nicholis, M.G., and M.J. Rutherford, "Experimental Constraints on Magma Ascent Rate for the Crater Flat Volcanic Zone Hawaiite," *Geology*, 32(6): 489–492 [2004].

NUCLEAR REGULATORY COMMISSION
ADVISORY COMMITTEE ON NUCLEAR WASTE
WASHINGTON, D.C. 20555-0001

November 3, 2004

The Honorable Nils J. Diaz
Chairman
U.S. Nuclear Regulatory Commission
Washington, D.C. 20555-0001

SUBJECT: THE 2005 RECOMMENDATIONS OF THE INTERNATIONAL COMMISSION ON
RADIOLOGICAL PROTECTION

Dear Chairman Diaz:

During its 154th meeting on October 19-21, 2004, the Advisory Committee on Nuclear Waste
(ACNW) held a working group meeting (WGM). The meeting included presentations from staff
and experts regarding the most recent draft recommendations of the International Commission
on Radiological Protection (ICRP). These draft recommendations were also presented to staff
and the public at NRC headquarters on September 15, 2004, by the Chairman of ICRP,
Dr. Roger Clarke, and the Vice Chairman (and Chairman-Elect) Dr. Lars-Erik Holm. The
Committee was represented at the September 15 presentations.

The ACNW WGM was held (1) to develop the information necessary to provide a letter report to
the Commission, (2) to understand the technical bases for the draft 2005 ICRP recommendations, (3) to review these recommendations against current NRC regulations and practice;
and (4) to identify aspects of the draft ICRP recommendations that may need further study. The
Committee heard presentations and discussions by:

Donald Cool, NRC staff and ICRP Committee 4 (practical applications); Vince Holahan, NRC
staff; Keith Eckerman, ORNL & ICRP Committee 2 (dosimetry); Michael Boyd, EPA; Edgar
Bailey, State of California and Conference of Radiation Control Program Directors; Richard
Vetter, Mayo Clinic and Advisory Committee on Medical Uses of Isotopes (ACMUI) member;
Dana Powers, member of the Advisory Committee on Reactor Safeguards (ACRS).

The draft ICRP recommendations cover eight areas:

Radiation quantities
Biological aspects
ICRP's general system of protection
ICRP's quantitative recommendations
Concepts of optimization
Exclusions from the recommendations
Medical exposures of patients as a separate issue
A proposal for protection of the environment

There are various incremental changes in the first seven areas, including radiation- and tissue-
weighting factors, new definitions for dosimetric quantities, and further discussion of the ICRP's
concepts of justification of practices, source constraints, and dose limits. An important point
about these draft recommendations is that ICRP's quantitative recommendations for workers

and members of the public have not changed since their 1990 recommendations, as published in ICRP Publication 60.[1] ICRP characterizes this update as a "simplification and elaboration" of its previous recommendations.

The eighth item is a proposal on radiological protection of non-human species. ICRP will form (mid-2005) a new committee to develop this proposal. The ACNW recommends that no action be taken at this time and that the NRC staff remain cognizant of the ICRP activities in this area until more details about ICRP's proposals are forthcoming. The Committee believes this is consistent with the Commission's documented direction to the staff.[2]

The Committee and the NRC staff cannot completely review the draft ICRP recommendations since the five comprehensive "foundation documents" (which give the scientific basis for the recommendations) are not yet available. Some of these foundation documents are expected soon. Others were reported by expert panel members to be still in progress.

The remainder of this letter concerns the draft ICRP recommendations in the first seven areas.

The unanimous view from expert panel members at the WGM was that there would likely be no significant improvement in the protection of worker and public health and safety by adopting these draft recommendations. Expert panel members identified potential difficulties, including confusion in the ICRP's use of terminology, confusion regarding ICRP's use of concepts such as safety culture without clear definition, and the application of the ICRP quantitative recommendations to U.S. licensees. Expert panel members did note that several elements of the recommendations would be improvements to the scientific basis. Some other elements need further consideration:

1. Without sufficient time to study and understand the foundation documents, it does not seem reasonable that the Draft ICRP recommendations should become final in June 2005. The Committee believes that the ICRP should allow more time for comment.

2. In its discussion of optimization, ICRP introduced the concept of "safety culture." It would be better if the ICRP specified the attributes of safety culture it finds important, rather than simply saying safety culture is part of optimization.

3. The Committee finds the current ICRP recommendations to be sufficient regarding "optimization." The Committee questions whether the draft ICRP recommendations are really improvements. ALARA as practiced in the U.S. provides a framework for accomplishing much of what the ICRP says about "optimization." ALARA is well understood and ALARA programs identify both dose reduction opportunities and other safety issues. The draft ICRP recommendations would unnecessarily complicate existing ALARA principles and applications with new terminology or dimensions.

---

[1]ICRP (1991) 1990 Recommendations of the International Commission on Radiological Protection. ICRP Publication 60. Ann. ICRP 21 (1-3), Pergamon Press, Oxford.

[2]Memorandum from A. Vietti-Cook, to W. Travers, EDO, "Staff Requirements - SECY-04-0055 - Plan for Evaluating Scientific Information and Radiation Protection Recommendations," ML041340304, May 13, 2004.

4. In the U.S. the term "best available technology" is a legal term and has ramifications that may not be consistent with ICRP objectives. ICRP should explain the application of "best available technology" within an optimization process for control of emissions to the environment.

5. In the U.S. there is a well-defined system of protection that is based on the relationship between radiation dose and risk. This relationship is not evident in the draft ICRP recommendations. The Committee believes that the draft ICRP recommendations would be improved by a detailed discussion of this relationship and its use in protecting the public.

6. The Committee believes that the ICRP goal of simplifying its terminology has not been achieved. For example, the term "constraint" in the draft ICRP recommendations has multiple meanings, some of which overlap with the meaning of the U.S. term "limit." The draft ICRP recommendations use the term "failure" to indicate not meeting a constraint. This may or may not mean that a legal or regulatory limit has been exceeded. These are examples of the confusion that can arise in trying to interpret and translate the terminology from the draft ICRP recommendations into practice.

RECOMMENDATIONS

The Committee believes that the Commission should consider deferring action on any of the draft ICRP recommendations until BEIR VII is published and available for review. Further, the expert panel members identified several items in the draft ICRP recommendations that could enhance the current regulations or radiation protection guidance. The Commission should consider three of these items as it deliberates on its response to the draft ICRP recommendations:

1. The radiation weighting factors for neutrons and protons (quality factors in 10 CFR Part 20)

2. The tissue-weighting factors that reflect the ICRP's current thinking about cancer risk

3. The ICRP's more recent methods and models for assessment of internal radiation exposures

Sincerely,

Michael T. Ryan
Chairman

March 11, 2005

The Honorable Nils J. Diaz
Chairman
U.S. Nuclear Regulatory Commission
Washington, D.C. 20555-0001

SUBJECT: STATUS OF THE AGREEMENT STATE PROGRAM AND THE INTEGRATED
MATERIALS PERFORMANCE EVALUATION PROGRAM (IMPEP)

Dear Chairman Diaz:

At its 156th Meeting, December 13-14, 2004, the NRC staff briefed the Advisory Committee on
Nuclear Waste (ACNW) on the status of the Agreement State Program. The Committee heard
details about the staff's use of the Integrated Materials Performance Evaluation Program
(IMPEP) to oversee and review the Agreement State Program. IMPEP results are used to
determine the adequacy and compatibility of individual Agreement State programs. IMPEP was
started in 1995 to replace the staff's previous prescriptive review program.

Currently there are 33 Agreement States regulating about 17,000 licensees. Two additional
States are pursuing agreements. Several other States have made inquiries asking for
background information about the Agreement State Program.

In the past 10 years, IMPEP has evolved in positive ways. Two new policy statements have
been issued (Statement of Principles and Policy for the Agreement State Program, Policy
Statement on Adequacy and Compatibility of Agreement State Programs). New procedures
have been developed for processing an agreement *(SA-700)*. A revised management directive
has been put in place for these reviews (Management Directive 5.6, "Integrated Materials
Performance Evaluation Program") and policy, rules, and guidance have been implemented
(Management Directives 5.3 and 6.3).

Furthermore, Agreement States are increasingly involved in administration of IMPEP and
NRC's materials program. Agreement State staff members have participated as members of:

- IMPEP review teams,
- Management Review Boards,
- The topical working groups, e.g., portable gauge rulemaking, working group and
  steering committee for materials security, and the national source tracking system, and
- Since 1997 they have taken a leadership role in the Organization of Agreement States.

Two key factors make the IMPEP program proactive rather than reactive and risk informed and
performance based rather than prescriptive. First, the collaboration of independent Agreement
State staff members and NRC's regional materials program staff on review teams provides for
consistency among the States and lets them share their results and experiences. This
interaction has led to improved risk-informed approaches and procedures.

13

Second, IMPEP ratings and responses use a graded approach with progressively more significant levels of action. The response levels go from Monitoring and Heightened Oversight (Procedure SA-122) to Probation (Procedure SA-113). NRC also can initiate an Emergency Suspension (Procedure SA-122), Suspension of an Agreement (Procedure SA-114), and Termination of an Agreement (Procedure SA-115). Future inspection frequency and the depth of interaction with Agreement State Program staff are determined by review of a program's performance. Additionally, the number of review team members is scaled to be proportional to the size of an Agreement State program. This graded approach allows for effective oversight and identification of Agreement State programs needing attention, so that corrective measures can be implemented before significant problems arise.

The staff told the Committee that the Agreement States face challenges in several areas:

- Integrating the regulation of Atomic Energy Act (AEA) materials with naturally occurring and accelerator-produced radioactive materials.

- Recruiting and retaining sufficient numbers of adequately trained staff to implement radiation protection programs.

- State financial constraints.

The Agreement State Program staff is aware of and effectively monitors these issues. The NRC staff tracks nuclear program events quarterly to identify emerging trends under these issues. The Committee will follow up on these problem areas in future briefings.

In summary, the Committee believes that Agreement State Program staff is providing effective and timely support to and oversight of individual Agreement State programs.

Sincerely,

Michael T. Ryan
Chairman

March 25, 2005

The Honorable Nils J. Diaz
Chairman
U.S. Nuclear Regulatory Commission
Washington, D.C. 20555-0001

SUBJECT: STATUS OF HIGH-SIGNIFICANCE AGREEMENTS ASSOCIATED WITH THE
             PROPOSED HIGH-LEVEL WASTE REPOSITORY AT YUCCA MOUNTAIN

Dear Chairman Diaz:

During the 157th meeting of the Advisory Committee on Nuclear Waste on February 23–25, 2005, the Committee was briefed by the NRC staff on the status of the key technical issue (KTI) agreements associated with the proposed high-level waste repository at Yucca Mountain. A total of 293 KTI agreements had been established to address data and analysis needs pertaining primarily to post-closure repository performance. As a result of these meetings and agreements, DOE committed to provide the information necessary to ensure a quality license application (LA) and efficient LA review by the NRC.

The Committee has been proactive with regard to the issue resolution process and related topics for several years. The Committee was briefed by DOE and NRC representatives during its 121st, 122nd, and 123rd meetings, September 19–21, October 17–19, and November 27–29, 2000, respectively, on progress toward resolution of KTIs (Reference 1). During its 133rd meeting March 19-21, 2002, the NRC staff briefed the Committee on the development of methods for performing sensitivity analyses as part of the total system performance assessment review (Reference 2). During its 143rd meeting June 24–25,2003, the NRC staff briefed the Committee on ranking agreements by risk significance and using risk information to resolve issues (Reference 3). The ACNW has also reported on other activities for risk-informing the issue resolution process (Reference 4).

At the 157th meeting, the staff informed the Committee that responses have been received from DOE for all 293 agreements, and reviews related to 224 agreements have been completed. Information concerning the remaining 69 agreements is currently under review. These reviews are expected to be completed by April 15, 2005.

According to the staff, most of the agreements, including the agreements currently under review, are of low or medium risk significance. The staff has identified only 41 high-significance agreements and finished reviewing the information on these agreements. Based on these reviews, the staff concluded DOE has fulfilled its obligation to provide information regarding 32 high-significance agreements. Resolution of most of the remaining high-significance agreements is not expected to be problematic as resolution of these agreements is pending DOE's release of information to the public and some model clarifications. The staff, however, has categorized a few high-significance agreements as "difficult issues," (e.g., agreements on volcanism and aircraft hazards).

The Committee offers the following comments and observations:

o   The staff noted that though agreements were "closed" at this pre-license application stage, any issue or topic would be fully evaluated during the review of a license application and that "closing" an agreement does not preclude additional review of an issue or topic after a license application is submitted.

o   The NRC staff's agreement resolution process has been efficient and risk-informed, and the staff has completed reviews in a timely but deliberate manner.

o   The pre-licensing technical exchanges and reviews have resulted in agreements on many technical issues. Other issues were identified as needing additional attention. The KTI resolution process should improve the quality of a potential DOE LA and the efficiency of the NRC staff's licensing review.

The Committee recommends that the staff continue using its pre-licensing KTI resolution process. In addition, because the KTI agreements are focused on the post-closure issues and only a small number of pre-closure issues were covered by the agreements, the Committee believes that the staff should also now focus on pre-closure issues. The Committee will proactively interact with the staff on the difficult issues that have been identified by the agreement resolution process, including issues associated with volcanism and aircraft hazards.

Sincerely,

Michael T. Ryan
Chairman

References:

1.   Letter dated February 8, 2001, from B. John Garrick, Chairman, Advisory Committee on Nuclear Waste, to Richard A. Meserve, Chairman, U.S. Nuclear Regulatory Commission, transmitting ACNW recommendations and concerns pertaining to the NRC high-level radiative waste issue resolution process. The letter is based on briefings by DOE and NRC representatives during the 121st, 122nd, and 123rd meetings of the Advisory Committee on Nuclear Waste September 19–21, October 17–19, and November 27–29, 2000, respectively, on progress toward resolution of the KTIs.

2.   Letter dated August 7, 2002, from George M. Hornberger , Chairman, Advisory Committee on Nuclear Waste, to Richard A. Meserve, Chairman, U.S. Nuclear Regulatory Commission, transmitting ACNW recommendations pertaining to parametric sensitivity and uncertainty analysis. The letter is based on briefings by NRC representatives during the 133rd meeting of the Advisory Committee on Nuclear Waste, March 19–21, 2002, on high level waste performance assessment sensitivity studies.

3.   Letter dated August 13, 2003, from B. John Garrick, Chairman, Advisory Committee on Nuclear Waste, to Richard A. Meserve, Chairman, U.S. Nuclear Regulatory Commission, transmitting ACNW comments including recommendations on the NRC staff's issue resolution process for risk-informing the sufficiency review of DOE's technical basis documents for the Yucca Mountain site recommendation.

4.   Letter dated September 28, 2001, from George M. Hornberger , Chairman, Advisory Committee on Nuclear Waste, to Richard A. Meserve, Chairman, U.S. Nuclear Regulatory Commission, transmitting ACNW comments and recommendations on the NRC staff's issue resolution process for risk-informing the NRC sufficiency review of DOE's technical basis documents for the Yucca Mountain site recommendation.

UNITED STATES
NUCLEAR REGULATORY COMMISSION
ADVISORY COMMITTEE ON NUCLEAR WASTE
WASHINGTON, D.C. 20555

April 27, 2005

The Honorable Nils J. Diaz
Chairman
U.S. Nuclear Regulatory Commission
Washington, DC 20555-0001

SUBJECT: BRIEFING ON RES-USDA RESEARCH: ESTIMATING GROUND
             WATER RECHARGE AND EVALUATING MODEL ABSTRACTION
             TECHNIQUES

Dear Chairman Diaz:

During its 158th meeting on March 15-17, 2005, the Advisory Committee on Nuclear
Waste heard presentations from the Office of Nuclear Regulatory Research and the
Agriculture Research Service of the U.S. Department of Agriculture (USDA) about their
research on estimating groundwater recharge and evaluating model abstraction
techniques. The main thrust of the research is to develop insights leading to (a) better
understanding of near surface water movement, saturated zone recharge, and solute
transport at sites with complex processes and features, and (b) guidelines on selecting
models that are as simple as possible but are realistic enough to provide a basis for
risk-informed decisionmaking. The Committee believes that this research should
continue.

The Committee learned about work that is being done to evaluate model abstractions of
subsurface water flux and pathways at a highly instrumented, densely sampled
watershed-scale site operated by the USDA in Beltsville, MD. This work builds on
earlier experiments conducted in well-controlled environments. Ground-penetrating
radar coupled with soil moisture measurements has been used successfully at the
Beltsville site to identify the location of preferred subsurface pathways that are
important to the assessment of uncertainty in infiltration and groundwater recharge
estimation.

This research shows:

•   Infiltration and groundwater recharge can be better understood using the
        methodology developed in this research. Models used to predict the fate and
        transport of contaminants in subsurface environments are sensitive to these
        parameters.

19

- These field tests can be used to evaluate alternative conceptual models and improve the selection of the best model abstraction.

- The Beltsville facility provides the opportunity for large-scale field testing in a highly instrumented environment. The research setting permits realistic estimates for sites similar in hydrology and subsurface geology to Beltsville through the incorporation of dynamic hydrologic processes.

The Committee offers the following conclusions and recommendations:

- Continued collaboration between the NRC and the USDA is a cost-effective way to participate in high quality research that is relevant to NRC needs. The Committee noted that the cost to NRC to date has been approximately 2% of the total cost.

- The Committee believes that this collaborative research program is important because it is aimed at reducing model complexity and assessing uncertainty while maintaining realism and the ability to support risk-informed decisionmaking.

- Both the field studies and the model abstraction research appear to have important applications in the areas of site characterization, flow and contaminant transport modeling, performance assessment, contaminant isolation technology evaluation, the design of monitoring programs, and uncertainty assessment.

- The Committee encourages the research staff to develop strategies to enable the transfer of results from studies at Beltsville to other hydrologic environments.

- The Committee believes the Beltsville research program should be coordinated with similar programs. For example, field-scale hydrologic research is being conducted at DOE facilities in Washington (Hanford) and New Mexico (Sandia) and at the University of Arizona's Maricopa site. Experience from these other sites should allow extension of the methodology developed in this research.

Sincerely,

Michael T. Ryan
Chairman

UNITED STATES
NUCLEAR REGULATORY COMMISSION
ADVISORY COMMITTEE ON NUCLEAR WASTE
WASHINGTON, D.C. 20555

June 21, 2005

The Honorable Nils J. Diaz
Chairman
U.S. Nuclear Regulatory Commission
Washington, DC 20555-0001

SUBJECT: PROPOSED RULE ON NATIONAL SOURCE TRACKING OF SEALED
            SOURCES

Dear Chairman Diaz:

During its 159th meeting on April 18-19, 2005, the Advisory Committee on Nuclear Waste
(ACNW) discussed a proposed rule that would require a national source tracking system for
sealed radioactive sources. The ACNW also discussed sealed source tracking and control
during its 156th meeting December 13-14, 2004. The Committee had the benefit of discussions
with the Department of Energy, the Maryland Department of the Environment/Radiological
Health Program, the Conference of Radiation Control Program Directors (CRCPD), and the
NRC staff.

The Committee commends the NRC staff for its leadership in creating a U.S. sealed source
tracking system. The system will focus on larger sources that pose greater risks. The
Committee believes that the tracking system, as currently envisioned, is appropriate. The
system requires the owners of large sources to register them when manufactured or received,
report changes of ownership when transfer is completed, and annually verify their inventory.

The Committee offers the following recommendations.

- The system is intended to operate online. As a consequence, care must be taken to
  ensure the tracking system remains secure from unauthorized entry while still being
  accessible to users.

- Information will be entered into the system by the manufacturers or owners of the sealed
  sources. The quality of the information entered into the system must be ensured.

- The Committee continues to see significant progress in the planning for control and
  tracking of sealed sources. While the proposed sealed source tracking system is an
  appropriate and useful first step, the Committee believes a continuing effort is needed to
  make the tracking system comprehensive, consistent, and risk informed. Federal and
  State agencies, the CRCPD, and the Organization of Agreement States should be
  encouraged to participate.

We look forward to continuing to work with the staff and the other interested parties as they develop a national source tracking system.

Sincerely,

Michael T. Ryan
Chairman

Reference:
E-mail to R. Major from M. Horn, dated 3/30/2005, Subject: Proposed Rule—National Source Tracking of Sealed Sources (RIN 3150-AH48), undated, official use only

June 28, 2005

The Honorable Nils J. Diaz
Chairman
U.S. Nuclear Regulatory Commission
Washington, D.C. 20555-0001

SUBJECT: DEPARTMENT OF ENERGY PLANS FOR TRANSPORTING SPENT NUCLEAR
FUEL AND HIGH-LEVEL RADIOACTIVE WASTE .

Dear Chairman Diaz:

At its 159th meeting on April 18-19, 2005, the Advisory Committee on Nuclear Waste (ACNW)
heard a presentation by Gary Lanthrum, Director of the Office of National Transportation (ONT)
of the Department of Energy (DOE).

Summary of the ONT Presentation

ONT will build and operate a system for transporting spent nuclear fuel (SNF) and high-level
radioactive waste (HLW) to a repository at Yucca Mountain, Nevada. ONT has been organized
around four project areas:

- Institutional
- Operational Infrastructure
- Fleet Acquisition
- Rail Through Nevada

The Institutional Project will collaborate with stakeholders to refine the transportation system as
it is developed. A key effort will be to develop policy and procedures for awarding grants under
Section 180(c) of the Nuclear Waste Policy Act to assist State, Tribal, and local emergency
response personnel in preparing for repository shipments and to develop information for the
public and interested stakeholders.

The Operational Infrastructure Project will define, develop, implement, and demonstrate the
operational infrastructure needed to support waste transportation from the utility and DOE
locations where the SNF and HLW are currently stored to the proposed Yucca Mountain
repository. The transportation infrastructure is intended to ensure optimal transportation from
the origin sites to Yucca Mountain, but optimization depends on other factors: maximal
utilization of existing casks and other facilities, as few shipments as possible, acceptable and
safe routes, and rapid transportation from each origin site. Optimization is complicated by the
uncertainty about when fuel of various types will be shipped.

The Fleet Acquisition Project will define the approach to purchasing transportation casks and
rolling stock to support transportation to the repository. The ONT's goal is to procure the
minimum suite of casks and undertake as few certifications as possible. Existing casks will be
used as much as possible, but DOE has found that the existing casks will fill only about 30% of
the need.

A rail line will be built in Nevada to connect the repository to an existing main rail line. ONT is preparing a rail alignment environmental impact statement (EIS) in accordance with the National Environmental Policy Act. DOE is evaluating the environmental inputs of a 318-to-344-mile-long corridor beginning in Caliente, Nevada. As a result of the scoping hearings on this EIS and the ensuing approximately 4000 comments to DOE, several additional routes are being considered for the proposed rail line.

DOE is also asking to be allowed to take credit for fuel burnup; i.e., to recognize that relatively high-burnup fuel has significantly less fissile content, and significantly more radionuclides that can poison the fission reaction than fresh fuel or low-burnup fuel. ONT said that there is little data on this topic in the U. S. The French have developed a considerable database and are working with the DOE. The chance of a criticality is significantly lowered in high-burnup fuel, and if this credit is allowed, the amount of SNF in a shipment can be increased. Without burnup credit, the space in some transportation casks could not be utilized fully. As the amount of SNF per shipment increases, the number of shipments needed decreases.

ACNW Observations

- The entire SNF transport system should be optimized from storage at the site of origin through transport, receipt, repackaging and emplacement in the drift. The transportation plan should be integrated with the strategy and plan for emplacing the waste in the repository.

- The DOE plan to try to obtain burnup credit — credit for reducing the risk of criticality in high burnup fuel — appears to be a wise move toward more realism in analysis of transportation of SNF and toward increased transportation efficiency.

ACNW Recommendation

NRC staff should consider allowing realistic burnup credit for cask certification.

Sincerely,

Michael T. Ryan
Chairman

24

UNITED STATES
NUCLEAR REGULATORY COMMISSION
ADVISORY COMMITTEE ON NUCLEAR WASTE
WASHINGTON, D.C. 20555

June 28, 2005

The Honorable Nils J. Diaz
Chairman
U.S. Nuclear Regulatory Commission
Washington, D.C. 20555-0001

SUBJECT: DEFINITION OF A TIMESPAN OF REGULATORY COMPLIANCE FOR A
GEOLOGICAL REPOSITORY AT YUCCA MOUNTAIN

Dear Chairman Diaz:

In a decision dated July 9, 2004, the U.S. Court of Appeals for the District of Columbia Circuit
ruled that the 10,000-year compliance period (hereafter the time period of compliance or TOC)
specified by the U.S. Environmental Protection Agency (EPA) for its Yucca Mountain site-
specific radiation standards at 40 CFR Part 197 violated Section 801 of the Energy Policy Act of
1992 (EnPA). It is unclear what changes will be made to Part 197 to address this ruling, but
such changes will require the Commission to modify the regulations in 10 CFR Part 63. The
Committee believes it may be useful for the Commission to consider previous ACNW advice on
TOC, as well as other views on TOC.

BACKGROUND

Before 1992 the generic radiation standards and implementing regulations for evaluating
geologic repository sites and licensing repository designs were given at 40 CFR Part 191 and
10 CFR Part 60. In 1992 Congress directed EPA and NRC to develop new radiation standards
and NRC to develop implementing regulations for licensing of the Yucca Mountain site. In
developing radiation standards, Congress directed EPA to contract with the National Academy
of Sciences (NAS) to advise EPA on the appropriate technical basis for public health and safety
standards for any Yucca Mountain repository.

On August 1, 1995, the NAS issued its report, "Technical Bases for Yucca Mountain Standards"
(the TYMS report). The NAS concluded there was "no scientific basis for limiting the time
period of the individual risk standard to 10,000 years or any other value." According to the
Academy, "compliance assessment is feasible for most physical and geologic aspects of
repository performance on the time scale of the long-term stability of the fundamental geologic
regime — a time scale that is on the order of one million years at Yucca Mountain." The
Academy also concluded that humans may not face peak radiation risks until tens to hundreds
of thousands of years after the disposal of wastes, "or even farther into the future." The
Academy thus recommended "that compliance assessment be conducted for the time when the
greatest risk occurs, within the limits imposed by the long-term stability of the geologic
environment."

After the Academy issued its findings and recommendations, EPA promulgated its draft Part 197 standards in which it proposed a 10,000-year TOC. In so doing, EPA requested comments on the reasonableness of adopting the NAS-recommended TOC or "some other approach in lieu of the 10,000-year compliance period," that EPA favored. During the public comment period, DOE and NRC went on record supporting the 10,000-year TOC while the State of Nevada proposed adopting a TOC extending to the time of projected peak dose, as NAS recommended. After reviewing the public comments, EPA promulgated its final rule adopting the 10,000-year TOC and in doing so expressed the view that NAS' TOC recommendation was "not practical for regulatory decision-making."

PAST ACNW ADVICE

The most recent ACNW views on TOC were given in two 1996 letters. The first letter gave background on defining a repository TOC, discussed related regulatory principles and selection criteria, and recommended a two-tiered approach to defining a TOC.[1] The second letter provided additional detail on the proposed two-tiered approach to addressing TOC issues.[2]

The first tier of the ACNW recommended approach was to define a quantitative dose limit for the reasonably maximally exposed individual (REMI) at a specific time for times on the order of several thousand years. The second tier was to qualitatively compare the peak dose and uncertainties of the dose standard. The Committee's recommendation did not require a quantitative measure of compliance at the TOC because of the uncertainties in defining future processes and events.

INTERNATIONAL APPROACHES

There is no international consensus on TOC among standard-setting bodies, regulators, and developers. This is not surprising considering the differences in national policies and the variations in design concepts and geologic settings. The attached table shows the variability of international TOC durations. Generally, a multitier approach to timeframes is used with a quantitative evaluation based on an early assessment of 1000 to 10,000 years and a longer, qualitative evaluation of a million years or longer, but there are many exceptions. Some countries, such as Germany, have not specified a TOC, but are considering the use of safety indicators with a qualitative assessment to a million years or more but no less than 10,000 years. Canada has specified a 10,000-year TOC and requires evaluation to an unspecified period beyond 10,000 years to show that there are no dramatic increases in dose in the post-TOC years.

Member countries of the International Atomic Energy Agency and the Nuclear Energy Agency (NEA) are participating in continuing activities to develop a consensus on using the results of performance assessments over long periods of time. Both organizations have recommended a tiered approach for evaluating repository performance. Deliberations on this issue continue. In the Fall of 2005, we expect to review a draft report on NEA's most recent workshop.

---

[1] ACNW letter report dated June 7, 1996, "Time Span for Compliance of Proposed Yucca Mountain HLW Repository."

[2] ACNW letter report dated November 14, 1996, "Road Map to ACNW's Recommendation for TOC."

PATH FORWARD

Upon the release of EPA's draft rulemaking for public comment, the Committee plans to review the draft regulation, meet with the NRC staff and stakeholders, and report its observations and recommendations to the Commission. The Committee also anticipates being briefed on the results of a 2005 NEA workshop. The briefing will be useful in the NRC's effort to help develop an international consensus on the use of long-timeframe performance assessment results.

In addition, the ACNW plans to hold a working group meeting in the Fall of 2005 on technical issues associated with long-timeframe performance assessments at Yucca Mountain. The Committee will report to the Commission on the results of this working group meeting.

Sincerely,

Michael T. Ryan
Chairman

Attachment:
As stated

INTERNATIONAL APPROACHES TO DEFINING
A REGULATORY TIME OF COMPLIANCE (TOC)

More than 30 countries have research and development programs for managing long-lived radioactive wastes in geologic repositories (Witherspoon and Bodvarsson, 2001[1]). Currently, there is no international consensus among standard-setting bodies, regulators, or developers in these countries on the time scale for evaluating the safety of geologic repositories. An effort is underway in the Nuclear Energy Agency to address this issue (by NEA's Integration Group for the Safety Case or IGSC). This Timescales Project has produced two reports so far. [2,3] A third "state-of-the-art report" is in preparation and will likely be published in 2006.[4]

The table below lists TOCs for 10 countries, including the United States, that have standards or guidance in place for evaluating the safety of long-lived radioactive waste repositories. A review of the literature indicates that several of these countries have TOCs that range from 1000 to 1,000,000 years. In some cases, there is no regulatory TOC cutoff and the calculations can be carried out to as long as 100 million years after facility closure. The technical bases for the specification of a particular TOC vary among developers and regulators. Cutoff times (i.e., the duration of the TOC) have been justified on the basis that (a) the relative hazard (toxicity) of spent nuclear fuel vs. a naturally occurring uranium ore body; (b) the potential for multiple peak doses to future receptors; and/or (c) intergenerational equity concerns. For the purposes of comparison, the table includes the three time frames selected by the International Atomic Energy Agency (IAEA) for an analysis of the use of repository safety indicators.

---

[1] P.A. Witherspoon and G.S. Bodvarsson (eds.), "Geological Challenges in Radioactive Waste Disposal – Third Worldwide Review," Berkeley, Ernest Orlando Lawrence Berkeley National Laboratory, LBNL-49767, December 2001.

[2] Nuclear Energy Agency, "The Handling of Timescales in Assessing Post-Closure Safety of Deep Geological Repositories, Proceedings of April 16-18, 2002, Workshop, Paris, France, Paris, Nuclear Energy Agency/Organization for Economic Cooperation and Development, 2002. Also see Nuclear Energy Agency, "Integration Group for the Safety Case (IGSC) Workshop on Handling of Time Scales Assessing Post-Closure Safety – Compilation of Abstracts," Paris, Nuclear Energy Agency/Organization for Economic Cooperation and Development, NEA/RWM/IGSC(2002)6, June 2002.

[3] Nuclear Energy Agency, "The Handling of Timescales in Assessing Post-Closure Safety – Lessons Learnt from the April 2002 Workshop in Paris France," Paris, Nuclear Energy Agency/Organization for Economic Cooperation and Development, NEA No. 4435, 2004.

[4] Belgium proposed the Timescales Project to NEA's IGSC. The purpose of the project is to produce a "state-of-the-art" report to document a consensus for cutting off performance assessment calculations at a specific time, if possible. Belgian officials believe that it would be helpful to be able to cite an international report with a recommendation and a technical basis for the recommendation. Although the NEA document has not been drafted, a 1 million-year cutoff is beginning to emerge as an informal consensus TOC based on discussions among the participants.

-1-

A related concern is to use performance assessment results in accounting for the uncertainties of analyses. Performance assessments in timespans of less than 100,000 years are generally considered more reliable. Longer term assessments (TOCs greater than 100,000 years) are generally considered less reliable because the uncertainties increase with time.

Regardless of the length of the specified TOC, there is a consensus among practitioners that a multitier approach should be used to judge repository performance, as noted in the table below. Performance assessments of TOCs of less than 100,000 years are generally more quantitative and TOCs of more than 100,000 years are generally more qualitative.

Country	TOC	Comments
BELGIUM	Not established yet	Safety demonstration analyses for at least **100,000,000 years**. [a]
CANADA [b]	≈ 10,000 years	Demonstrate repository safety quantitatively with detailed calculations.
	< 100,000 years	Qualitative demonstration, using "reasoned arguments," that there is no dramatic increase in releases from repository after the first **10,000 years**.
	< 1,000,000 years [c]	An example for the purposes of the environmental impact statement to demonstrate that the radiological toxicity of spent fuel is equivalent to a natural uranium ore body.
FINLAND [d]	≈ 10,000 years	Evaluate repository performance over an environmentally predictable period.
	> 10,000 years	Do a stylized, quantitative calculation using a broad range of safety indicators.
	> 1,000,000 years	Do a qualitative calculation.
FRANCE [b, e]	0–500 years [f]	Do analysis for assumed period of passive institutional controls.
	< 50,000 years	Minimum period of environmental predictability. Analysis not intended to reflect future climate change and the onset of glaciation.
	> 50,000 years	Do a qualitative analysis as a reference, taking into account the expected evolution of repository system.
GERMANY	No specified time	Evaluate repository performance up to about **10,000 years**, taking into account period during which repository barriers would be subject to minor changes. [b]

Country	TOC	Comments
		Do an analysis on the order of **1,000,000 years** to identify repository sites with overall favorable geologic characteristics. [e] Do other demonstration analyses for beyond **1,000,000 years**. [g]
JAPAN [b]	**Not established yet**	Evaluate repository performance taking into account period of peak dose up to about **100,000,000 years**.[h]
SPAIN [i]	**Not established yet** (To be defined by 2010.)	Demonstration analysis to stop at **1,000,000 years**.
SWEDEN [j]	**< 1000 years**	Do a quantitative calculation.
	< 100,000 years[k]	Do a quantitative analysis, taking into account the next major glacial period. The analysis period must be greater than **10,000 years**.
	> 100,000 years	Do a stylized, qualitative calculation. The analysis is to stop at **1,000,000 years**.
SWITZERLAND [b]	**No specified time**	Duration for demonstration analysis terminated at **10,000,000 years**.[l]
UNITED KINGDOM [b]	**Not established yet**	Timeframe for analysis implied to be less than **1,000,000 years**.
UNITED STATES	**10,000 years** [m]	Timeframe for analysis for evaluation of transuranic (TRU) radioactive wastes.
	1,000,000 years[n, o]	Evaluate Yucca Mountain repository performance, taking into account periods of peak dose up to about **1,000,000 years**.
IAEA[p]	**< 10,000 years**	Quantitative analysis assuming the current biosphere and institutional controls.
	< 1,000,000 years	Mix of qualitative and quantitative "illustrative" calculations intended to reflect future climate change and the present-day reference biosphere
	> 1,000,000 years	Qualitative analysis during the period over which radiological toxicity of repository is equivalent to a natural uranium ore body.

REFERENCES:

[a] Studiecentrum voor Kernenergie – Centre d'étude de l'Energie Nucléaire (SCK·CEN – Belgian Nuclear Research Centre), "Identifying and Testing Indicators for Assessing the Long Term Performance of Geological Disposal Systems: The [European] SPIN Project, *SCK·CEN Scientific Report 2002*, Mol, Belgium, [2002].

-3-

[b]   U.S. General Accounting Office, "Nuclear Waste – Foreign Countries' Approaches to High-Level Waste Storage and Disposal," Washington, DC, Resources, Community and Economic Development Division, GAO/RCED-94-172, August 1994.

[c]   Atomic Energy of Canada Limited, "Environmental Impact Statement on the Concept for Disposal of Canada's Nuclear Fuel Waste," Mississauga, Ontario, AECL-10711, COG-93-1, September 1994.

[d]   The Radiation Protection and Nuclear Safety Authorities in Denmark, Finland, Iceland, Norway, and Sweden, "Disposal of High Level Radioactive Waste – Consideration of Some Basic Criteria," Stockholm, Sweden, Swedish Radiation Protection Institute (Statens Strålskyddsinstitut – SSI), 1993.

[e]   Committee on a Site Selection Procedure for Repository Sites (Arbeitskreis Auswahlverfahren Englagerstandorte – AkEnd), "Site Selection Procedure for Repository Sites: Recommendations of the AkEnd," Berlin, German Federal Ministry for the Environment, Nature Conservation, and Nuclear Safety, December 2002.

[f]   U.S. Environmental Protection Agency, "Spent Nuclear Fuel and High-Level Waste Programs Disposal Programs in Other Countries (Chapter 3) in Environmental Radiation Protection Standards for Yucca Mountain, Nevada – Draft Background Information Document for Proposed 40 CFR 197, "Office of Radiation and Indoor Air, EPA 402-R-99-008, August 1999.

[g]   Bundesanstalt für Feowissenschaften und Rohstoffe, "Grundsätze der Endlagerung radioaktiver Abfälle – Die Sicherheitsphilosophie des Bundesamtes für Strahlenschutz (Standards for the Permanant Disposal Site for Radioactive Waste – Safety Philosophy of the Federal Office of Radiation Protection), Salzgitter, German Federal Republic, 2004.

[h]   Japan Nuclear Cycle Development Institute [JNC], "H12: Project to Establish the Scientific and Technical Basis for HLW Disposal in Japan – Supporting Report 3: Safety Assessment of the Geological Disposal System," Ibaraki, Japan, Report TN1410 2000-004, 2000. [NOTE: Because of the current regulatory mandate to address international practices and standards, the Japanese are actively participating in the NEA Timescales Project.]

[i]   A. Astudillo, "Geological Disposal of High-Level Radioactive Wastes in Spain," in P.A. Witherspoon and G.S. Bodvarsson (eds.) Geological Challenges in Radioactive Waste Disposal – Third Worldwide Review, Ernest Orlando Lawrence Berkeley National Laboratory, LBNL-49767, December 2001.

[j]   Swedish Radiation Protection Institute, "Health, Environment and Nuclear Waste, SSI's Regulations and Comments," Stockholm, Sweden, SSI Report 99:22, 1999.

[k]   Swedish Nuclear Power Inspectorate (Statens Kärnkraftinspektion), the repository developer whose implementing recommendations, including a time scale for the analysis, will be defended at the time of licensing.

[l] National Cooperative for Radioactive Waste (National Genossenschaft für die Lagerung radioaktiver Abfälle – Nagra), "Project Opalinus Clay: Safety Report – Demonstration of Disposal Feasibility for Spent Fuel, Vitrified High-Level Waste and Long-Lived Intermediate-Level Waste (*Entsorgungsnachweis*)," Wettingen, Switzerland, Nagra Technical Report NTB 02-05, 2002. (Although the demonstration calculations were carried out to 10 million years, the Nagra report notes that there is little confidence in the calculations beyond 1 million years.)

[m] U.S. Environmental Protection Agency, "40 CFR Part 191: Environmental Standards for the Management of Spent Nuclear Fuel, High-Level and Transuranic Radioactive Wastes; Final Rule," *Federal Register*, Vol. 50, No. 182, pp. 38066-38089, September 19, 1985.

[n] National Research Council, "Technical Bases for Yucca Mountain Standards," Washington, DC, Commission on Geosciences, Environment, and Resources, National Academy Press, July 1995.

[o] U.S. Department of Energy, "Final Environmental Impact Statement for a Geologic Repository for the Disposal of Spent Nuclear Fuel and High-Level Radioactive Waste at Yucca Mountain, Nye County, Nevada, Vol. 1, Impact Analyses, Chapters 1 through 15, "Office of Civilian Radioactive Waste Management, DOE/EIS-0250, February 2002.

[p] International Atomic Energy Agency, "Safety Indicators in Different Time Frames for the Safety Assessment of Underground Radioactive Waste Repositories. First Report of the INWAC Subgroup on Principles and Criteria for Radioactive Waste Disposal," Vienna, Austria, IAEA-TECDOC-767, October 1994. (The NEA suggestion to evaluate until the dose from the spent fuel is equivalent to a uranium ore body would not likely require calculation beyond a million years.)

-5-

33

UNITED STATES
NUCLEAR REGULATORY COMMISSION
ADVISORY COMMITTEE ON NUCLEAR WASTE
WASHINGTON, D.C. 20555-0001

July 1, 2005

The Honorable Nils J. Diaz
Chairman
U.S. Nuclear Regulatory Commission
Washington, DC 20555-001

SUBJECT: COMMENTS ON ICRP FOUNDATION DOCUMENTS – A FOLLOWUP TO THE
            ACNW'S NOVEMBER 3, 2004 COMMENTS

Dear Chairman Diaz:

The ACNW has reviewed the five "Foundation Documents" offered by the International
Commission on Radiological Protection (ICRP) in support of its 2005 Draft Recommendations.
By this letter the ACNW reaffirms the recommendations in our November 3, 2004 letter and in
the March 16, 2005 briefing to the Commission. Nothing in the Foundation Documents
changes our earlier observations and recommendations.

As the ACNW stated, the Commission should consider deferring action on any of the Draft
ICRP Recommendations until BEIR VII is published and available for review, and consider
implementing changes in tissue weighting factors, radiation weighting factors, and more recent
methods and models for assessment of internal dose. There is no urgent need to make these
changes; they can be made when regulations are revised for other reasons.

The ACNW has several observations on the Foundation Documents:

1.   As written the Foundation Document on the "Representative Individual" lacks clarity.
      Even though it usefully clarifies compliance with dose limits (constraints); the term
      "representative individual" is used in different senses in the document. The definitions
      and their applications need to be clarified. Examples could be used to convey the intent
      and use of the various dose assessment protocols and strategies discussed in the
      document.

2.   Unless substantial clarifications are made to the definition and use of the "representative
      individual" concept, it offers little use when compared to the concepts of the "Average
      Member of a Critical Group" or the "Reasonable Maximally Exposed Individual" (RMEI).

3.   Consistent with its November 3, 2004 letter, the ACNW recommends that the
      Commission defer consideration of the Foundation Documents regarding the "Biology"
      and "Dosimetry" until the BEIR VII Committee report is issued and available for review
      and comparison.

4. The ACNW believes that the additional guidance provided in the Foundation Document on "Optimization" would not substantially improve current ALARA programs, or protection of workers, the public, or the environment. The principle of stakeholder involvement discussed in the Optimization document is consistent with the Commission's current programs and activities as discussed in the agency's Strategic Plan and implementing documents.

5. Regarding the draft Foundation Document on "The Concept and Use of Reference Animals and Plants for the Purposes of Environmental Protection," the ACNW continues to hold the view expressed during our March 16, 2005 briefing to the Commission: that there has been no evidence to contradict the philosophy that by protecting humans the environment is protected. This Foundation Document tries to make the case that separate recommendations are needed or justified.

More detailed comments are given on the foundation documents in the Attachment.

Sincerely,

Michael T. Ryan
Chairman

Attachment: Detailed comments on the
ICRP Foundation Documents

ATTACHMENT: DETAILED COMMENTS ON THE ICRP FOUNDATION DOCUMENTS

Foundation Document "Assessing Dose of the Representative Individual for the Purpose of Radiation Protection of the Public"

The document is very repetitive. Basic concepts, ideas, and approaches are repeated many times. Unfortunately, terms like "representative individual" are slightly different in each instance. The Abstract, Executive Summary, and Introduction all cover the same ground with different terminology.

The value of the document is derived from its focus on several principles:

1. Both nonstochastic (deterministic) and stochastic assessments have a place. The document offers comments on where each is best employed. The document should be more focused on this point. Clear examples should be given for each case and the limitations should be spelled out. A common criticism of nonstochastic analysis is that true risk can be missed. ICRP should offer a case to counter this assertion.

2. For nonstochastic assessments, doses below a limit ("constraint" in ICRP terminology) demonstrate compliance. This is a helpful statement.

3. For probabilistic risk assessment, the document suggests compliance with a dose limit: if the 95th percentile of the dose distribution is within a factor of 3, compliance is demonstrated. This needs clarification. Additionally, the ICRP should advise regulators on how to make the compliance algorithm clear. Examples would help to demonstrate these concepts.

Major drawbacks to the document are:

The "representative individual," as presented in the document, is discussed in contradictory ways. Paragraph 23 states:

> Therefore, for the purpose of protection of the public, it is necessary to characterize an individual, either hypothetical or specific, who receives the highest dose which can be used for determining compliance with the dose constraint. This individual is defined as the representative individual.

How can a representative individual get the "highest dose?"

Paragraph (S9) uses a slightly different definition:

> The representative individual is the hypothetical individual receiving a dose that is representative of the most highly exposed individuals in the population.

This definition implies that the representative individual is a member (perhaps the average, median, or mode) of the most highly exposed group. This qualitative definition is subject to interpretation and is not consistent with paragraph 23.

-1-

Paragraphs 67 and 68 imply that the representative individual possessed "mean" characteristics regarding habits that are not "outside the range of day-to-day life." This is not easily reconciled with the individual who receives "the highest dose." ICRP needs to clarify the definition and guidance.

Temporal uncertainty and variability seem not to have been considered. It appears that the approaches to dose calculations address only uncertainty and variability in spatial data. This report seems to indicate that once determined (for a specific point in time), the parameters used to model pathways of exposure and calculated dose are fixed throughout the entire life span of the exposed individual. The dose calculations need to consider temporal uncertainty and variability over time. Both are known to be important.

Foundation Document "Biological and Epidemiological Information on Health Risks Attributable to Ionising Radiation: A Summary of Judgements for the Purposes of Radiological Protection of Humans"

1. This Foundation Document suggests small adjustments to "detriment adjusted nominal probability coefficients for cancer." These small adjustments do not substantially change previous cancer risk values. In addition, additional analyses are expected in the Biological Effects of Ionizing Radiation Committee of the National Academy of Science Report (BEIR VII), expected later this year. The ACNW continues to believe that the Commission should consider deferring action on any of the draft ICRP recommendations until the BEIR VII Report is published and available for review.

2. A related finding is reported: "For cancer and hereditary disease at low doses/dose rates the use of a simple proportionate relationship between increments of dose and increased risk is a scientifically plausible assumption." This conclusion further supports taking no action until the BEIR VII report is published. ICRP recommends no large changes in risk factors.

3. The Foundation Document states: "Knowledge of the roles of induced genomic instability, bystander cell signaling and adaptive response in the genesis of radiation–induced health effects is insufficiently well developed for radiological protection purposes; in many circumstances these cellular processes will be incorporated in epidemiological measures of risk." The ACNW believes that this statement is a fair assessment of the state of knowledge of these issues at this time though new information is reported regularly. The ACNW will keep informed of newer studies and report to the Commission as appropriate.

4. The document states: "Proposed changes in radiation weighting factors for protons and neutrons are noted; these judgements are fully developed in the ICRP Committee 2 Foundation Document, Basis for dosimetric quantities used in radiological protection (FD-C-2)". This additional report provides substantive detail. The Foundation Document on "Biological and Epidemiological Information..." states that: "New radiation detriment values and tissue weighting factors have been proposed; the most significant changes from ICRP 60 relate to breast, gonads and treatment of remainder tissues." ACNW's comments on FD-C-2 are provided separately below.

-2-

Foundation Document "Draft for Discussion International Commission on Radiological Protection Committee 2 Basis for Dosimetric Quantities Used In Radiological Protection"

The two principal recommendations in this report are to change the radiation weighting factors for protons and neutrons and change the tissue weighting factors used to calculate the effective dose (formerly referred to as dose equivalent).

For protons, the ICRP recommends that the weighting factor be lowered from 5 (the value recommended in ICRP Publication 60[1]) to 2. Currently, in 10 CFR 20.1004, Table 1004(B).1, Quality Factors and Absorbed Dose Equivalencies, a quality factor of 10 is given for high energy protons. Consistent with our letter of November 3, 2004, the ACNW believes that the Commission should consider updating this quality factor, but that the update can be done by issuing regulatory guidance or at a time when the regulations are revised for other reasons. The ICRP has developed a method to calculate the quality factor for neutrons as a function of neutron energies. Three equations for three different energy ranges are recommended in Equation 4.7:

$$W_R = \begin{cases} 2.5 + 18.2\,e^{-[\ln(E_n)]^2/6} & , \quad E_n < 1\,\text{MeV} \\ 5.0 + 17.0\,e^{-[\ln(2E_n)]^2/6} & , \quad 1\,\text{MeV} \le E_n \le 50\,\text{MeV} \\ 2.5 + 3.25\,e^{-[\ln(0.04E_n)]^2/6} & , \quad E_n < 50\,\text{MeV} \end{cases} \qquad (4.7)$$

Neutron energy (MeV) (thermal).....	Quality factor (Q) 10 CFR 20.1004 (B) 2	Values Calculated from New ICRP Methods	Ratio of ICRP Recommended Value to Current 10 CFR 20.1104
2.50E-08	2	2.5	1.25
1.00E-07	2	2.5	1.25
1.00E-06	2	2.5	1.25
1.00E-05	2	2.5	1.25
1.00E-04	2	2.5	1.25
1.00E-03	2	2.5	1.25
1.00E-02	2.5	3.0	1.21
1.00E-01	7.5	10.0	1.34
5.00E-01	11	19.3	1.75
1	11	22.0	2.00
2.5	9	19.8	2.20

The table above shows that the current quality factors for neutrons differ from those using the ICRP's recommended formulas by factors ranging from 1.21 to 2.20. These factors are not substantially different and given the uncertainties in determining neutron spectra in practical

---

[1] ICRP. 1990 Recommendations of the ICRP. ICRP Publication 60. *Ann of the ICRP*, **21**(1-3). Pergamon Press, Oxford (1991).

radiation protection situations, these factors may often be comparable to the errors associated with such measurements. Consistent with its letter of November 3, 2004, the ACNW believes that the Commission should consider incorporating this method of calculating neutron quality factors, but that the update can be done through regulatory guidance or at a time when the regulations are revised for other reasons.

This Foundation Document, along with the Foundation Document on Biological and Epidemiological Information, also suggests changes to tissue weighting factors:

> "In the proposals for the new Recommendations the W_T for remainder (0.12) is divided equally between the 15 specified tissues given in Table 2, i.e. approximately 0.008 each. This value is smaller than the least value assigned to any of the named tissues (0.01). In practice this gives the arithmetic average of the doses to these 15 tissues. Since the formulation of remainder is the same in every case the system preserves additivity in effective doses which is a considerable advantage in practical radiation protection."

This change clarifies how to calculate dose to other organs not specifically assigned weighting factors.

In changing these weighting factors, to be consistent it would be necessary to recalculate the existing Annual Limits on Intake and Derived Air Concentrations used in current regulations.

"The Optimisation of Radiological Protection - Broadening the Process," Report by the ICRP Committee 4 Task Group on Optimisation of Protection

The ACNW observed in its letter of November 3, 2004, that

> "current ICRP recommendations ....[are] sufficient regarding "optimization." The Committee questions whether the draft ICRP recommendations are really improvements. ALARA as practiced in the U.S. provides a framework for accomplishing much of what the ICRP says about "optimization." ALARA is well understood and ALARA programs identify both dose reduction opportunities and other safety issues. The draft ICRP recommendations would unnecessarily complicate existing ALARA principles and applications with new terminology or dimensions."

The ACNW believes the additional guidance provided in this Foundation Document would not substantially improve current ALARA programs or protection of workers, the public, and the environment.

Additionally, this Foundation Document provides ICRP's views on the "role of the stakeholder." The ACNW believes that the Commission has developed significant initiatives to involve stakeholders in the regulatory process as described in the Strategic

-4-

Plan and implementing documents and programs, particularly with regard to "openness" [reference: NRC's Strategic Plan: FY 2000 - FY 2005, NUREG-1614, Vol. 2, part 1].

Foundation Document: "The Concept and Use of Reference Animals and Plants for the Purposes of Environmental Protection"

The ACNW believes that the ICRP has failed to make a case for overturning the principle that has guided radiation protection practice for more than 50 years. This principle states that protecting humans also protects the environment. The ICRP says in paragraph (6):

> The Commission [ICRP] still believes that this judgement is likely to be correct in general terms, because the steps taken to protect the public, by reference to dose limits for them, have resulted in strict controls and limitations on the quantities of radionuclides deliberately introduced into the environment."

The ACNW believes that the ICRP has not provided any evidence to contradict this long-standing principle.

Further, it seems clear that the ICRP's guidance is driven by other concerns. As the ICRP states:

> However, there are now other demands upon regulators, in particular the need to comply with the requirements of legislation directly aimed at the protection of wildlife and natural habitats; the need to make environmental impact assessments with respect to the environment generally; and the need to harmonise approaches to industrial regulation, bearing in mind that releases of chemicals from other industries are often based upon their potential impact upon both humans and wildlife.

The ACNW believes that this ICRP recommendation goes far beyond radiation protection issues and is more relevant to strategies for national policy on radiation protection. It is telling that in the last quote the ICRP cites "chemicals from other industries" as an example but does not explain why radioactive materials should be included with chemicals. The justification for this linkage is not clear and in any case is not developed or substantiated in the text.

July 27, 2005

The Honorable Nils J. Diaz
Chairman
U.S. Nuclear Regulatory Commission
Washington, D.C. 20555-0001

SUBJECT: RESPONSE TO THE OCCUPATIONAL SAFETY AND HEALTH AGENCY
REQUEST FOR INFORMATION ON IONIZING RADIATION

Dear Chairman Diaz:

On May 3, 2005, the Occupational Safety and Health Agency (OSHA) submitted the following request for information in the *Federal Register*:

> OSHA requests data, information and comment on issues related to the increasing use of ionizing radiation in the workplace and potential worker exposure to it. Specifically, OSHA requests data and information about the sources and uses of ionizing radiation in workplaces today, current employee exposure levels, and adverse health effects associated with ionizing radiation exposure. OSHA also requests data and information about practices and programs employers are using to control employee exposure, such as exposure assessment and monitoring methods, control methods, employee training, and medical surveillance. The Agency will use the data and information it receives to determine what action, if any, is necessary to address worker exposure to occupational ionizing radiation.

The Advisory Committee on Nuclear Waste (ACNW or Committee) considered OSHA's request for information (RFI) as published in the *Federal Register* and is providing its independent views on OSHA's RFI.

The Committee notes that many components of a robust system of radiation protection, including radiation protection programs, regulations and regulatory agreements, and other sources of information, already exist:

1.   NRC and Agreement States regulations promulgated for activities regulated by the Atomic Energy Act (AEA);

2.   State radiation protection programs for non-AEA radioactive materials;

3.   Federal guidance on sources of electronic product radiation from the Center for Devices and Radiological Health of the Food and Drug Administration;

4.   State programs for electronic product radiation control;

5.   U.S. Environmental Protection Agency general applicable radiation protection statutes and related guidance;

6. U.S. Department of Energy radiation protection statutes (10 CFR Part 835, "Occupational Radiation Protection"), regulations, orders, and guidance;

7. Reports of the National Academy of Sciences, including the recent report "Health Risks from Exposure to Low Levels of Ionizing Radiation" Biological Effects of Ionizing Radiation (BEIR) VII - Phase 2, 2005;

8. The Conference of Radiation Control Program Directors (CRCPD) and the Organization of Agreement States (OAS) programs that support Agreement State and non-Agreement State radiation protection programs;

9. The CRCPD and OAS joint letter to OSHA regarding its RFI;

10. NRC data on occupational radiation exposure (NUREG-0713, Volume 25, "Occupational Radiation Exposure at Commercial Nuclear Power Reactors and Other Facilities," 2003);

11. Nuclear Energy Institute (NEI) data on occupational radiation exposure;

12. Guidance offered by the National Council on Radiation Protection and Measurements (NCRP); and

13. OSHA-NRC Memoranda of Understanding.

This information demonstrates that existing programs provide adequate radiation protection to workers. We have summarized some of the information in the appendix to this letter.

The Committee also believes that the premise of OSHA's request for information that worker exposure might be increasing is not substantiated. For example, the ACNW notes that in Table 3.1 of NUREG-0713 (see the appendix to this letter), the trend in average measurable Total Effective Dose Equivalent (TEDE) per worker has decreased in every one of the six categories of NRC licensees (from 1994 to 2003).

The Committee did not have access to any comprehensive database for radiation dose information for radiation workers in medical areas that use non-AEA radioactive materials or electronic product radiation devices and cannot comment on trends for these workers. The ACNW notes that these workers' groups are monitored under State authority. The 33 Agreement States typically integrate these non-AEA radiation worker monitoring and protection programs into NRC-approved programs. Nonetheless, the ACNW cannot include this radiation worker group in the remaining comments in this letter.

The NEI provided additional analysis to the Committee indicating a clear trend in worker dose reduction in the nuclear power industry for collective dose per reactor and collective dose per megawatt year of operation. The NEI data on average annual number of workers with measurable dose for the period of 1973 - 2003 show a decreasing trend since 1984. The NEI reported that these trends are a result of robust As Low As Reasonably Achievable (ALARA) programs rather than a focus only on strict numerical standards. The ACNW interprets the data to indicate that the current limits, along with the implementation of the ALARA principle, have been effective in providing radiation protection for workers.

While collective dose for Department of Energy (DOE) workers has increased from 2002 - 2003, this increase reflects more work activities rather than an increase for individual workers (DOE/EH-0688, "DOE Occupational Radiation Exposure 2003 Report").

Moreover, the recently released BEIR VII report affirms that cancer risk estimates for exposure to ionizing radiation have not changed significantly from those reported in previous BEIR reports.

In summary, the ACNW believes that existing radiation safety programs and the current regulatory infrastructure promote effective and timely oversight of occupational radiation protection programs required under Federal and State authorities. Furthermore, documented trends in worker exposures do not support the need for a new regulatory initiative. The ACNW recommends that the Commission provide a response to OSHA consistent with this view.

Sincerely,

Michael T. Ryan
Chairman

APPENDIX
INFORMATION EVALUATED BY ACNW REGARDING OSHA'S
REQUEST FOR INFORMATION (RFI)

NRC Data on Occupational Radiation Exposure

NRC summarizes information regarding worker exposure from its databases for several industry segments. The latest available summaries are provided in NUREG-0713, Volume 25, "Occupational Radiation Exposure at Commercial Nuclear Power Reactors and Other Facilities." Example data from NUREG-0713 (Table 3.1) are provided below. The table shows the average annual exposure for certain categories of NRC licensees: namely industrial radiography, manufacturing and distribution, low-level waste disposal, independent spent fuel storage, fuel cycle licenses, and commercial light water reactors. The table indicates a downward trend in the collective dose (person-rem) from 1994 to 2003 across the industries measured. This observation further supports ACNW's view that the system of radiation protection is robust and effective; thus, OSHA need not intervene to address worker exposure to occupational ionizing radiation.

Agreement State Programs

In its recent review of the NRC Agreement States program, the ACNW found the radiation program to be robust and effective in providing radiation protection programs for workers regulated under both Atomic Energy Act (AEA) and non-AEA-regulatory authority. ACNW has reported previously on the Integrated Material Performance Evaluation Program (IMPEP), created to oversee and review the Agreement States program. IMPEP results are used to determine the adequacy and compatibility of individual Agreement State programs. In the ACNW's 2005 letter to the Commission, "Status of the Agreement State Program and the Integrated Materials Performance Evaluation Program (IMPEP)," the Committee stated the following:

> Two key factors make the IMPEP program proactive rather than reactive, and risk informed and performance based rather than prescriptive. First, the collaboration of independent Agreement State staff members and NRC's regional materials program staff on review teams provides for consistency among the States and lets them share their results and experiences. This interaction has led to improved risk-informed approaches and procedures. Second, IMPEP ratings and responses use a graded approach with progressively more significant levels of action.

> Future inspection frequency and the depth of interaction with Agreement States Program staff are determined by review of a program's performance.

> This graded approach allows for effective oversight and identification of Agreement State programs needing attention, so that corrective measures can be implemented before significant problems arise.

TABLE 3.1
Average Annual Exposure Data for Certain Categories of NRC Licensees
1994-2003

NRC License Category* and Program Code	Calendar Year	Number of Licensees Reporting	Number of Monitored Individuals	Number of Workers With Measurable TEDE	Collective TEDE (person-rem)	Average TEDE (rem)	Average Measurable TEDE per Worker (rem)
Industrial	1994	139	2,886	2,007	1,415	0.49	0.71
Radiography	1995	149	3,761	2,651	1,443	0.38	0.54
	1996	148	3,766	2,639	1,449	0.38	0.55
03310	1997	148	3,570	2,574	1,356	0.38	0.53
03320	1998	142	4,952	3,446	1,863	0.38	0.54
	1999	132	3,837	2,827	1,551	0.40	0.55
	2000	129	3,368	2,542	1,528	0.45	0.60
	2001	124	3,780	3,161	2,111	0.56	0.67
	2002	100	3,420	2,842	1,729	0.51	0.61
	2003	86	2,649	2,319	1,504	0.57	0.65
Manufacturing	1994	44	2,941	1,251	580	0.20	0.46
and	1995	36	2,666	1,222	595	0.22	0.49
Distribution	1996	38	2,631	1,241	556	0.21	0.45
	1997	33	1,154	665	397	0.34	0.60
02500	1998	31	1,986	654	402	0.20	0.61
03211	1999	39	2,181	836	419	0.19	0.50
03212	2000	39	2,481	1,188	415	0.17	0.35
03214	2001	36	1,862	1,211	351	0.19	0.29
	2002	29	1,437	1,052	328	0.23	0.31
	2003	23	1,849	1,459	394	0.21	0.27
Low-Level	1994	2	202	83	22	0.11	0.27
Waste	1995	2	212	56	8	0.04	0.15
Disposal**	1996	2	165	67	8	0.05	0.12
	1997	2	185	50	5	0.03	0.11
03231	1998	1	27	13	1	0.05	0.10
	1999	0					
Independent	1994	1	158	89	42	0.27	0.47
Spent Fuel	1995	1	104	49	51	0.49	1.04
Storage	1996	1	97	53	54	0.56	1.02
	1997	1	55	24	6	0.11	0.24
23100	1998	1	53	21	3	0.05	0.12
23200	1999	2	86	33	5	0.06	0.16
	2000	2	146	83	6	0.04	0.07
	2001	2	154	107	13	0.08	0.12
	2002	2	75	67	6	0.08	0.09
	2003	2	55	46	3	0.05	0.06
Fuel	1994	8	3,596	2,847	1,147	0.32	0.40
Cycle	1995	8	4,106	2,959	1,217	0.30	0.41
Licenses -	1996	8	4,369	3,061	878	0.20	0.29
Fabrication	1997	10	11,214	3,910	1,006	0.09	0.26
Processing and	1998	10	10,684	3,613	950	0.09	0.26
Uranium Enrich.	1999	9	9,693	3,927	1,020	0.11	0.26
	2000	9	9,336	4,649	1,339	0.14	0.29
21200	2001	9	8,145	3,980	1,162	0.14	0.29
21210	2002	8	7,937	3,886	661	0.08	0.17
	2003	8	7,738	3,633	556	0.07	0.15
Commercial	1994	109	139,390	71,613	21,704	0.16	0.30
Light Water	1995	109	132,266	70,821	21,688	0.16	0.31
Reactors***	1996	109	126,402	68,305	18,883	0.15	0.28
	1997	109	126,781	68,372	17,149	0.14	0.25
41111	1998	105	114,367	57,466	13,187	0.12	0.23
	1999	104	114,154	59,216	13,666	0.12	0.23
	2000	104	110,557	57,233	12,652	0.11	0.22
	2001	104	104,928	52,292	11,109	0.11	0.21
	2002	104	107,900	54,460	12,126	0.11	0.22
	2003	104	109,990	55,967	11,956	0.11	0.21
Grand Totals	1994	303	149,173	77,890	24,910	0.17	0.32
and Averages	1995	305	143,115	77,758	25,003	0.17	0.32
	1996	306	137,430	75,366	21,828	0.16	0.29
	1997	303	142,959	75,595	19,919	0.14	0.26
	1998	290	132,069	65,213	16,406	0.12	0.25
	1999	286	129,951	66,839	16,661	0.13	0.25
	2000	283	125,868	65,695	15,940	0.13	0.24
	2001	275	118,869	60,751	14,746	0.12	0.24
	2002	243	120,769	62,307	14,850	0.12	0.24
	2003	223	122,281	63,424	14,413	0.12	0.23

* These categories consist only of NRC licensees. Agreement State licensed organizations are not required to report occupational exposure data to the NRC.

** As of 1999, there are no longer any NRC licensees involved in this activity. All low-level waste disposal facilities are now located in Agreement States and no longer report to the NRC.

*** Includes all LWRs in commercial operation for a full year for each of the years indicated. Reactor data have been corrected to account for the multiple counting of transient reactor workers (see Section 5).

Environmental Protection Agency (EPA) Radiation Protection Programs and Requirements

The EPA has responsibility for protecting the public with considerable authority for developing radiation protection program guidance and setting environmental standards. The EPA has wide-ranging authority to promote, conduct, or contract research for radiation protection information; to promulgate generally applicable environment standards which limit man-made radioactive materials; to provide technical assistance to the States and other Federal agencies with radiation protection programs; to advise them in the execution of such programs; and to provide emergency assistance in responding to radiological emergencies. While EPA's generally applicable radiation protection standards apply to protection of members of the public, they are coordinated with requirements promulgated by NRC and the States.

Department of Energy (DOE) Radiation Protection Programs and Requirements

The DOE's 10 CFR 835, "Occupational Radiation Protection," provides nuclear safety requirements that, if violated, provide a basis for the assessment of civil and criminal penalties. The DOE has a series of guides, standards, programs, and orders which are consistent with 10 CFR 835. The DOE's Office of Health and Safety establishes comprehensive and integrated programs for the protection of workers from hazards in the workplace, including ionizing radiation. The DOE has standard radiation dose limits which establish maximum permissible doses to workers and members of the public. DOE radiation protection standards are based on EPA 1987 guidance, which in turn is based on recommendations from the International Commission on Radiological Protection (1977) and the National Council on Radiation Protection and Measurements (NCRP) (1987). In addition to the requirement that radiation doses not exceed the limits, contractors are required to maintain ALARA exposures.

According to DOE/EH-0688, "DOE Occupational Radiation Exposure 2003 Report,"

> The change in operational status of DOE facilities has had the largest impact on radiation exposure over the past 5 years due to the shift in mission from production to cleanup activities and the shutdown of certain facilities. For 2003, this resulted in an increase in the collective dose as sites handled more radioactive materials for processing, storage, or shipping.

In this document, DOE also stated that a statistical analysis of data over the past 5 years indicates "that while the collective TEDE, neutron, and extremity dose increased between 2002 to 2003, it does not represent a statistically significant change in the dose received by individual workers at DOE."

Other Data Sources

ACNW considered several databases:

- Specific information related to incidents in Agreement and non-Agreement States was included from the NRC's nuclear materials events database (NMED), *http://www.nmed.inl.gov.*

- State radiation control programs most often integrate regulation and control of ionizing radiation and radioactive material not regulated by NRC under the Atomic Energy Act

-2-

(as amended). Sources of information include the Conference of Radiation Control Program Directors (CRCPD) <http://www.crcpd.org> and the Organization of Agreement States (OAS) <http://www.agreementstates.org>.

- Recent examples of emerging guidance include: the work cosponsored by the Center for Devices and Radiological Health (CDRH) and the Transportation Security Administration (TSA) and performed by the NCRP. This work is reported in the "Presidential Report on Radiation Protection and Advice: Screening of Humans for Security Purposes Using Ionizing Radiation Scanning Systems." The report will be completed and delivered to CDRH this summer. The CDRH intends to use the NCRP recommendations as guidance when considering new performance standards. The CDRH also is working with other government agencies and the American National Standards Institute Committee (ANSI) N43 to identify new consensus standards for cargo and vehicle scanners that use ionizing radiation.

- The National Academy of Sciences recently released its BEIR VII report "Health Risks from Exposure to Low Levels of Ionizing Radiation," which provides an update to health risks related to radiation. The report affirms that current cancer risk estimates have not changed significantly from earlier estimates.

- The 2003 DOE Occupational Radiation Exposure Report provides a summary and analysis of the occupational radiation exposure received by individuals associated with DOE activities.

OSHA-NRC Memoranda of Understanding

There are four Memoranda of Understanding (MOU) between OSHA and NRC.

1. **STD 01-04-001 – STD 1-4.1 OSHA Coverage of Ionizing Radiation Sources Not Covered by the Atomic Energy Act 10-30-1978.** This early memorandum recognizes the U.S. Atomic Energy Commission (AEC) authority to regulate source, by-product, and certain special nuclear materials, and that OSHA's authority to regulate radiation sources does not include those regulated by AEC. It further states that OSHA covers all radiation sources not regulated by AEC, such as X-ray equipment, accelerators, accelerator-produced materials, electron microscopes, betatrons, and some naturally occurring radioactive materials.

2. **CPL 02-00-086 – CPL2.86 – Memorandum of Understanding Between OSHA and NRC.** This memorandum characterizes NRC-licensed nuclear facility hazards into four categories:

- Radiation hazards produced by radioactive materials;

- Chemical hazards produced by radioactive materials;

- Plant conditions which affect the safety of radioactive materials and thus present an increased radiation hazard to workers; and

-3-

- Plant conditions which result in occupational hazards, but do not affect the safety of the licensed radioactive materials.

This MOU delineates the general areas of responsibility of each agency, describes generally the efforts of the agencies to achieve worker protection at facilities licensed by NRC, and provides guidelines for coordination of interface activities between OSHA and NRC. To insure against gaps in the protection of workers and avoid duplication of effort, the MOU acknowledges NRC jurisdiction over the first three hazards and OSHA over the fourth hazard.

3. **Worker Protection at Facilities Licensed by the NRC 11-16-1998.** This MOU describes the efforts of the agencies to achieve worker protection at facilities licensed by NRC and provides guidelines for coordination of interface activities between OSHA and NRC. The accord replaced existing guidelines which had been used to coordinate activities of the two agencies. OSHA will provide NRC information, based on reports of injuries or complaints, about nuclear power plant sites where increased management attention to worker safety is needed. OSHA also will give training in basic chemical and industrial safety to NRC inspection personnel so that they will be able to better identify matters of concern to OSHA in radiological and nuclear inspections. The NRC will provide training in radiation safety to those OSHA and State program personnel who may participate in joint evaluation of safety hazards in some facilities.

4. **Gaseous Diffusion Plant Sites.** The AEA, as amended, created the United States Enrichment Corporation (USEC), to manage and operate the two uranium gaseous diffusion enrichment plants in Paducah, Kentucky, and Piketon, Ohio. The AEA requires USEC to be subject to and comply with the Occupational Safety and Health Act, and with applicable NRC standards for radiological safety and common defense and security. Furthermore, the USEC Privatization Act requires NRC and the OSHA to enter into a memorandum of agreement to coordinate their regulatory programs to assure worker safety, avoid regulatory gaps in the protection of workers, and avoid duplicative regulation.

-4-

UNITED STATES
NUCLEAR REGULATORY COMMISSION
ADVISORY COMMITTEE ON NUCLEAR WASTE
WASHINGTON, D.C. 20555-0001

August 3, 2005

The Honorable Nils J. Diaz
Chairman
U.S. Nuclear Regulatory Commission
Washington, D.C. 20555-0001

SUBJECT: REPORT ON SELECTED NRC-SPONSORED TECHNICAL ASSISTANCE
PROGRAMS AT THE CENTER FOR NUCLEAR WASTE REGULATORY
ANALYSES

Dear Chairman Diaz:

During the past 16 months the Advisory Committee on Nuclear Waste (ACNW) has written five
letters to the Commission describing results of the ACNW's continuing oversight of the Nuclear
Regulatory Commission's (NRC's) regulatory technical assistance and research programs. The
topics discussed were selected programs of the Center for Nuclear Waste Regulatory Analyses
(CNWRA) (March 4, 2004), radionuclide transport (May 5, 2004), uranium dioxide solubility
(July 6, 2004), model uncertainty (August 4, 2004), and groundwater recharge (April 27, 2005).
The ACNW also briefed the Commission on the research program on March 16, 2005.

As part of the Committee oversight, three members of the ACNW led a focused discussion of
selected technical assistance topics on April 13-15, 2005 at the CNWRA in San Antonio, Texas.
Two ACNW consultants supported these members. The Technical Director of the CNWRA had
previously provided the ACNW team an overview of the accomplishments of the CNWRA and
future projects during the 157th ACNW meeting in February 2005. The team focused its April
2005 discussions on activities addressing topics likely to be important in evaluating a license
application for a potential repository at the Yucca Mountain site and of particular interest to the
ACNW.

This letter, the first of two addressing topics discussed during the April 2005 visit, deals with the
CNWRA work on corrosion, radionuclide mobility, and performance assessment modeling. A
second letter will address analysis of a potential igneous event at Yucca Mountain and its
possible consequences.

Summary of the team's Yucca Mountain-related observations:

(1)   The presentations on container life, the radionuclide source term, the near-field
        environment, radionuclide retardation, and the published versions of the Department of
        Energy's Total System Performance Assessment were comprehensive and illustrated
        the strength of the CNWRA in these areas.

(2)   The CNWRA has made significant progress in ongoing work directed at understanding
        the controls and the processes involved in container corrosion. Laboratory corrosion
        studies include stress corrosion cracking resistance of Alloy 22, high-level waste glass
        dissolution processes, mechanical properties of the waste package, and the relationship
        between in-package chemistry and package corrosion. The laboratory studies show
        that corrosion by chloride-containing solutions can be inhibited by appropriate ratios of
        certain anion concentrations. Studies of Yucca Mountain dust within the waste
        emplacement drifts indicate that nitrate and sulfate are present in sufficient

concentration to potentially inhibit corrosion. The results of corrosion rate studies are expressed as distributions that incorporate uncertainty in corrosion rates. The CNWRA's humidity deliquescence studies show that, although chloride deliquescence could form corrosive brine, other components of this dust can inhibit such corrosion. The CNWRA is abstracting these results for incorporation in the ongoing model development activities.

(3)   Regarding spent fuel dissolution studies in support of the total-system performance assessment, the CNWRA staff is using parameter values from the technical literature and results from laboratory experiments to model the dissolution of radionuclides from spent fuel. These studies have shown that fuel burnup does not significantly influence dissolution of the uranium dioxide matrix.

(4)   The CNWRA has been responsive to the suggestions made during the ACNW's Geosphere Transport Working Group meeting (ACNW letter to Chairman Diaz dated August 3, 2004). Potential spatial water chemistry impacts on sorption have been evaluated. Additional experiments are underway to determine neptunium sorption in the alluvium. Retardation in the alluvium can provide a barrier to radionuclide migration, and understanding the spatial variability of retardation reduces uncertainty.

(5)   The CNWRA is currently evaluating improvements in the modeling of phenomena such as tephra remobilization, consequences of drift degradation, drip shield and waste container weld corrosion, and colloid transport. Furthermore, numerous parameter values and their distributions reflect recent progress in the understanding of relevant features, events, and processes (FEPs). This work is ongoing and is expected to lead to improvements in evaluation of the risk associated with the FEPs involved in the performance of the proposed repository.

(6)   The CNWRA has ongoing programs that address the frequency, consequences, and potential health effects that are associated with igneous activity, and will publish a number of letters in the next several months. The ACNW will continue to interact with the NRC staff on this subject and will provide a letter to the Commission in the near future.

The CNWRA reported to the ACNW team on its evaluation of models and codes for use in pathway dose assessment for complex decommissioning applications and expects to complete a final letter in October 2005. The ACNW plans to review this work when it is completed.

The ACNW will continue its dialog and meetings with the NRC and CNWRA staffs and will keep the Commission apprised of our view of the progress of this work.

Sincerely,

Michael T. Ryan
Chairman

UNITED STATES
NUCLEAR REGULATORY COMMISSION
ADVISORY COMMITTEE ON NUCLEAR WASTE
WASHINGTON, D.C. 20555-0001

August 12, 2005

The Honorable Nils J. Diaz
Chairman
U.S. Nuclear Regulatory Commission
Washington, D.C. 20555-0001

Dear Chairman Diaz:

SUBJECT: DRAFT REVISED DECOMMISSIONING GUIDANCE TO IMPLEMENT THE
LICENSE TERMINATION RULE

The NRC staff is developing revised decommissioning guidance to implement the License
Termination Rule (LTR). In support of this effort, NRC staff and the ACNW (the Committee)
have participated in two meetings. The first was an April 2005 decommissioning workshop
organized by the NRC staff. The entire Committee attended this workshop. The second was a
1-day working group meeting on June 15, 2005, during the 160th meeting of the Committee.

In its working group meeting, the Committee had the benefit of discussions with the NRC staff
and five invited experts selected to provide the perspective of experienced practitioners.[1]
During the meeting, the Committee provided comments and suggestions that the staff is
considering while developing the draft guidance. Since the staff participated in the working
group meeting and subsequent Committee deliberations, the Committee is confident that its
comments and suggestions have been conveyed.

The working group discussed a range of guidance revisions in several different areas. The
Committee has not seen the revised document since it is still being developed. However,
observations and recommendations that have been discussed with the staff are provided in the
rest of this letter.

OBSERVATIONS AND RECOMMENDATIONS

- The Committee supports the issuance of generic guidance implementing the LTR.
  However, site-specific factors are especially important to consideration of partial
  restricted release under the long-term control (LTC) license and intentional soil mixing.

---

[1] The invited experts were Eric Abelquist, Director of the Radiological Assessments and Training
Program, Oak Ridge Institute for Science and Education; Virgil Autry, Consultant, Department of Health
and Environmental Control, State of South Carolina; Eric Darois, Radiation Safety and Control Services in
New Hampshire; Tracy Ikenberry, Associate and Senior Health Physicist, Dade Moeller & Associates; and
Thomas Nauman, Vice President, Shaw Environmental and Infrastructure.

In these cases, the Committee recommends that the NRC staff develop criteria and a demonstration process to enable site-specific decisions on a case-by-case basis.[2]

- The staff presented an approach to classifying restricted-use sites as either lower or higher risk and a graded approach to selecting institutional controls. The Committee believes that this approach is appropriate and risk informed.

- Durable controls will be needed for higher risk restricted-use sites. NRC staff reported that the guidance will provide two options: an LTC license and a legal agreement/restrictive covenant (LA/RC) with the NRC. The second option, while potentially attractive to a site owner, may present uncertainties with respect to the survivability of the long-term controls. The staff prefers the LTC approach, and the Committee concurs with this preference.

- The staff asked the Committee for its input on the merits of partial restricted release. The staff indicated a preference for including the entire site under the LTC license, and the Committee agrees. However, there may be site-specific factors that merit consideration, and the Committee recommends a case-by-case approach to partial restricted release.

- Existing guidance on the use of engineered barriers is limited. The Committee concurs with the staff's assessment that the agency needs expanded generic guidance on the barrier design options and more performance experience that can be tailored to specific sites. The breadth and depth of this guidance should be sufficient to enable risk-informed decisionmaking.

- The staff prefers robust engineered barriers. However, the experience base for the performance of currently favored designs goes back only a few decades. Very long-term performance (centuries to millennia) has not yet been demonstrated, and there is no basis for concluding that current systems will perform for very long times without continuing periodic maintenance. The Committee concurs with the staff's assessment that monitoring will be needed to confirm performance.

- The Committee recommends that the conventional upper bound resident farmer scenario be used only as a screening tool and that realistic scenarios be used to evaluate risk. The revised guidance will address the use of more realistic scenarios for projected land use. Many decommissioning sites can achieve unrestricted release using the very conservative and unrealistic resident farmer scenario, but guidance is needed on more realistic exposure scenarios, especially for complex materials sites.

---

[2] The ACNW recommended a case-by-case approach to requests for intentional mixing of contaminated soil in its letter of July 30, 2004, "Review of the LTR Analysis - Intentional Mixing of Contaminated Soil." The Committee notes that the working group expert panel was divided with respect to the merits of permitting intentional mixing of contaminated soils.

- Groundwater monitoring should be a prime consideration in the revised guidance and should address ways to determine the requirements for subsurface characterization and monitoring. The guidance should also address subsurface characterization, monitoring plans, and contingency plans should groundwater contamination occur.

- The Committee recognizes that the lessons learned from decommissioning projects provide valuable information for designing new facilities (designing with the end in mind). In addition to developing protocols and mechanisms for information collection and dissemination, the staff will need to devise a process to evaluate the accuracy and reliability of the information that is disseminated.

The Committee has participated in the staff's information-gathering activities for the revised decommissioning guidance to be published at the end of September 2005. Therefore, the staff need not respond to the issues discussed in this letter. The Committee has discussed these issues with the staff and plans to interact with the staff again after the draft guidance is published. The Committee believes that these early and ongoing interactions have helped the Committee and the staff meet their respective obligations on schedule.

The Committee plans to comment on the draft guidance when it is published.

Sincerely,

Michael T. Ryan
Chairman

UNITED STATES
NUCLEAR REGULATORY COMMISSION
ADVISORY COMMITTEE ON NUCLEAR WASTE
WASHINGTON, D.C. 20555-0001

September 29, 2005

The Honorable Nils J. Diaz
Chairman
U.S. Nuclear Regulatory Commission
Washington, D.C. 20555-0001

SUBJECT: REVIEW OF STAFF'S PRECLOSURE REVIEW PREPARATIONS FOR THE
            PROPOSED YUCCA MOUNTAIN REPOSITORY

Dear Chairman Diaz:

At its 162nd meeting on August 2-4, 2005, the Advisory Committee on Nuclear Waste (ACNW) heard a presentation by the Nuclear Regulatory Commission (NRC) staff on "Status of Yucca Mountain Preclosure Review Preparations." The following are our observations and recommendations regarding the staff's preparations to meet the challenge of this risk-informed, performance-based review.

Background

The NRC staff has undertaken a number of activities to prepare for its review of preclosure design aspects of the license application for the proposed Yucca Mountain repository. An important part of these activities is the organization of review teams for performance assessment, engineering, site characterization, and health physics. The staff is also developing a list of risk-significant topics based on the staff's experience and analysis and on information obtained from visits to relevant fuel-handling facilities. The staff is concentrating on high-risk topics (including the related uncertainties) and on structures, systems, and components that can prevent or mitigate the impacts of postulated event sequences.

The staff identified topics for detailed prelicensing review. The topics include aircraft crash hazard and event sequences, criticality and seismic event sequences, and preclosure safety analysis. The staff has begun to discuss these topics in a series of technical exchanges with the U.S. Department of Energy (DOE).

Observations

The NRC staff informed the Committee that preclosure design aspects of licensing are receiving increased attention and that the staff is applying necessary resources to address them. The staff is developing guidelines for staff interaction with DOE on preclosure topics before the license application submittal. The Committee agrees with this approach. However, the Committee recognizes that the efficiency and effectiveness of the staff's efforts have been challenged by the apparent lack of completeness and detail in available information on the design of preclosure systems, processes, facilities, and equipment that are important to operational safety.

59

The Committee concurs that the staff's initial list of review topics is appropriate for evaluation. Additional topics for evaluation are identified in periodic staff meetings. The Committee believes iterative preclosure safety assessments and relevant licensing experience (e.g., Private Fuel Storage) are potentially useful in identifying additional topics. The rigor of the staff's approach to preparing and modifying the list of preclosure focus topics would be easier to recognize if the staff had a documented basis for the choice of topics (and, as appropriate, a basis for exclusion of topics).

The Committee believes that lessons learned from other nuclear regulatory licensing experience could also be a useful source of topics for the staff's preclosure review. For example, human reliability and fire protection may dominate the risk at both reactor and nonreactor facilities if not considered early in the design stage. Risk insights indicate that without attention to human reliability aspects of design and adequate training in the early stages of design, human error can be a significant contributor to accidents associated with movement of heavy loads at reactor facilities. A significant number of heavy-load lifts, load manipulations, and movements are expected to occur during the preclosure operational stage of the repository. They should therefore be evaluated in the preclosure review. Likewise, costly fire protection retrofitting at reactor facilities occurred in the past because designers did not have a thorough understanding early in the design stage of the risk from fire. Fire and smoke propagation can lead to adverse system interactions and common-cause failures that may compromise multiple safety barriers.

Another topic deserving attention is equipment and facility aging analysis. The staff informed the Committee that it plans to consider aging effects in estimating the probability of failure of equipment. Given the lengthy period of operation that the DOE contemplates for the preclosure facility, these effects could be significant, although difficult to quantify. The Committee also notes that reliability goals for important preclosure equipment such as have been established for safety-significant reactor equipment could be a significant enhancement to preclosure safety.

Recommendations

1. The NRC staff should develop a documented, risk-informed process for identifying topics that the staff will focus on in reviewing preclosure aspects of the proposed Yucca Mountain repository. Iterative safety assessments could be a useful tool in such a process.

2. The staff should add human reliability analysis and fire protection to the list of high-priority preclosure review topics.

3. The staff should assess DOE's reliability targets for systems and components important to safety and those factors that impact reliability during the preclosure period (e.g., design configuration, operation, equipment and facility aging, surveillance, and maintenance).

4. To increase the efficiency and effectiveness of its preparations for a risk-informed performance-based review, the staff should continue to seek detailed information from DOE on preclosure design.

We look forward to hearing from the staff again on the subject of preclosure safety assessment at a mutually convenient future date.

Sincerely,

Michael T. Ryan
Chairman

UNITED STATES
NUCLEAR REGULATORY COMMISSION
ADVISORY COMMITTEE ON NUCLEAR WASTE
WASHINGTON, D.C. 20555-0001

September 30, 2005

The Honorable Nils J. Diaz
Chairman
U.S. Nuclear Regulatory Commission
Washington, D.C. 20555-0001

SUBJECT: COMMENTS ON USNRC STAFF RECOMMENDATION OF THE USE OF
COLLECTIVE DOSE

Dear Chairman Diaz:

On July 20, 2005, staff from the Office of Nuclear Regulatory Research briefed the Committee regarding proposals on effective and realistic uses of the concept of collective dose in radiation dose analysis.

The staff reported they are considering four options (one with three variations) regarding the uses of collective dose. These options as reported are as follows.

Option 1-Truncate individual doses at some nominal value.

- Truncate individual doses at some nominal value from the collective dose calculation.

- Truncate individual doses at some distance from a facility or at some future time.

Option 2-Health Physics Society position on collective dose

- For populations in which *almost* all individuals are estimated to receive a lifetime dose of less than 10 rem above background, collective dose is a highly speculative and an uncertain measure of risk and should not be used for the purpose of estimating population health risks [Radiation Risk in Perspective (position statement of the Health Physics Society),1996, revised in 2004].

- Estimation of health risk associated with radiation doses that are of similar magnitude as those received from natural sources should be strictly qualitative and encompass a range of hypothetical health outcomes, including the possibility of no adverse health effects at such low levels.

Option 3-Individual dose emphasis

- Emphasizes protection of individuals in the critical group of an exposed population and assumes that if the average individual in the critical group is protected, the entire population is protected.

- Consistent with the 10 CFR Part 20 Subpart E, "Radiological Criteria for License Termination Rule," which explicitly states that the average individual of the critical group must be below a 25 mrem per year dose constraint and ALARA.

63

- No collective dose is calculated in this option.

Option 4-Significance determination of a collective dose calculation

- Use a Commission-approved criterion to judge the significance of a collective dose calculation.

Option 4a: 1 mrem per year and 100 person-rem per year

- International bodies argue that it is not cost-beneficial to do a formal cost-benefit analysis process when individual and collective doses are less than 1 mrem per year and 100 person-rem per year, respectively, and the practice can be exempted from regulatory oversight (IAEA 1996, ICRP 1992, EC 1999).

Option 4b: Background collective radiation dose comparison

- Compare the collective dose from a regulated activity to the collective dose from background radiation to the same population.

- This approach is comparable to the approach in NUREG-1515, "Standard Review Plans for Environmental Reviews for Nuclear Power Plants."

Option 4c: Safety goal evaluation

- Expand the use of the reactor safety goal/quantitative health objective value for latent cancer fatalities of "0.1% of the sum of cancer fatality risks resulting from all other causes" to other applications that use collective dose.

- The staff would compare collective dose calculations to this safety goal value, either in units of person-rem or in latent cancer fatality risk, and make a determination of "not a significant additional risk."

Observations and Recommendations

The Committee believes that collective dose has little value in an *absolute* sense. Irrespective of whether very low doses can be reliably measured or estimated, the product of an individual dose and a population magnitude does not yield a number that has any real meaning. When estimates of risk are desired, the Committee recommends use of individual risk within the context of the critical group or the reasonably maximally exposed individual (RMEI) scenario.

However, the Committee does believe that collective dose is useful for comparing different management options (e.g., steps taken under ALARA to reduce radiation doses to workers).

The Committee believes that there is no basis for truncating dose at some nominal value when calculating collective dose.

Given the inherent limitations of collective dose and the serious potential for misuse (e.g., using collective dose as a measure of risk), the Committee does not recommend adoption of any of the options considered above.

Sincerely,

Michael T. Ryan
Chairman

UNITED STATES
NUCLEAR REGULATORY COMMISSION
ADVISORY COMMITTEE ON NUCLEAR WASTE
WASHINGTON, D.C. 20555-0001

October 27, 2005

The Honorable Nils J. Diaz
Chairman
U.S. Nuclear Regulatory Commission
Washington, DC 20555-0001

SUBJECT: PROJECT PLAN FOR THE YUCCA MOUNTAIN LICENSE APPLICATION
REVIEW BY NRC STAFF

Dear Chairman Diaz:

The NRC staff briefed the Committee regarding the development of a project plan for the Yucca Mountain license application review on September 20, 2005. The Division of High Level Waste Repository Safety staff provided an overview of the project plan, including the project management approach and the license application review process. The staff indicated that NRC will use robust and sound project management practices and will leverage best practices from other NRC licensing programs. The project plan includes cost, schedule, and control capabilities of information management tools aimed at supporting a defensible licensing process.

The Committee believes that the project plan will be effective in supporting a potential license application review and will provide timely and quality information to NRC management regarding:

- Work breakdown structures,
- Integrated schedule,
- Resource planning and management,
- Project risk management,
- Change assessment and management,
- Communications,
- Records management, and
- Performance measures.

The Committee believes the project plan will be valuable to license application reviewers from NRC and the Center for Nuclear Waste Regulatory Analyses. The staff developed the project plan from the requirements set forth in applicable statutes and regulations, as well as the Yucca Mountain Review Plan. The staff considered the schedule and content requirements for the Safety Evaluation Report. The staff also anticipates using the project plan to support Atomic Safety and Licensing Board hearings and other needs.

The Committee agrees with the staff that there is a need to develop performance measures for monitoring and assessing plan implementation. The staff also has developed a number of contingencies to address potential problems such as loss of key technical staff or technical issues arising during review, and the staff has a clear strategy and approach for managing the review of a potential license application.

The Committee believes that the staff is well positioned to begin the review of a potential license application and to keep management informed in a specific and timely way regarding the progress of the review.

Sincerely,

Michael T. Ryan
Chairman

UNITED STATES
NUCLEAR REGULATORY COMMISSION
ADVISORY COMMITTEE ON NUCLEAR WASTE
WASHINGTON, D.C. 20555-0001

December 9, 2005

The Honorable Nils J. Diaz
Chairman
U.S. Nuclear Regulatory Commission
Washington, DC 20555-0001

SUBJECT: REVIEW OF THE NRC PROGRAM ON THE RISK FROM IGNEOUS ACTIVITY
             AT THE PROPOSED YUCCA MOUNTAIN REPOSITORY

Dear Chairman Diaz:

The Advisory Committee on Nuclear Waste (the Committee) has met several times to discuss
the risk from igneous activity at the proposed Yucca Mountain repository. In September 2004,
the ACNW held a Working Group Meeting on this topic, and summarized its conclusions and
recommendations in a November 4, 2004 letter report. The Center for Nuclear Waste
Regulatory Analyses (CNWRA) staff updated members of the Committee on NRC's current
studies on volcanism in April 2005.

Subsequent meetings, review, and analysis of recently published documents of the NRC and its
contractors, and discussions with the NRC and CNWRA staffs have resulted in the following
observations and recommendations regarding potential igneous activity at the repository.
Several of the ACNW's observations and recommendations are related to the NRC staff's use
of assumptions in their analysis that appear to be conservative rather than realistic. Excessive
conservatism can foster misperceptions of the performance of the proposed Yucca Mountain
repository and conceal attributes of processes that should receive the attention of the NRC
staff. The Committee believes continued investigation of potential scenarios will better prepare
the staff to evaluate assumptions and approaches in a potential license application. The
Committee looks forward to understanding how the staff has used risk-informed thinking
throughout the analysis of igneous activity at the proposed Yucca Mountain repository.

INTERACTION BETWEEN INTRUDING MAGMA AND REPOSITORY DRIFT AND WASTE PACKAGES

The Committee believes that resolution of questions about the interaction between intruding
magma and the repository drift and waste packages could be better risk informed by
considering alternative interaction scenarios and their potential influence on consequences.
Specifically, the effects on repository performance of rapid magma cooling with attendant
increases in viscosity and solidification of magma should be considered in analyzing the
magma/drift/waste package interactions in scenarios in which the intruding dike vents to the
surface as a volcano. The alternative scenarios and their implications include the following:

1. Magma characteristics influence production of different materials when an igneous intrusion intersects a repository drift. If the volatile content of magma is relatively large, as anticipated from available evidence, volcanic ash could erupt into the drift at the point of dike/drift intersection. Only after the entrained gases have escaped from the magma due to eruption processes would magma enter the repository drift as a lava flow rather than as ash. The Committee has been provided information that either ash or lava will likely solidify near the entry point into the repository drift.

2. Key factors in the rate of solidification of the magma and self-sealing of the drift are the delivery rate of magma, latent heat of crystallization, volatile content of the magma, and thermal conductivity of drift walls and waste packages. As a result, the magma would likely interact with a few waste packages near the point of entry. Rapid solidification would likely prevent the formation of secondary (flank) vents from the rising magma flowing into repository drifts and subsequently venting to the surface.

3. Solidification of magma entering a repository drift is an important topic to consider regarding the integrity of waste packages. At present, both the Department of Energy (DOE) and the NRC staff assume that the contents of a relatively small number of waste packages directly involved in the dike intrusion are completely destroyed by interaction with invading magma and that all the included waste is entrained in the magma and becomes airborne after eruption. In contrast, Electric Power Research Institute (EPRI) modeling indicates that waste packages are sufficiently robust that invading magma will not destroy the packages (EPRI, 2004). Information presented to the Committee suggests that quenching of magma on an intact canister could provide a protective barrier, thereby isolating and protecting the waste from the intruding magma. Thus, even if a few waste packages are entrained within a cone-forming volcanic conduit, the NRC staff's alternative approach that assumes complete destruction of the waste canisters may lead to incorrect assumptions and parameterization in performance assessment. Undue conservatism also may lead to a distorted view of the risks posed by the repository.

4. Waste packages will be most resistant to degradation and therefore to igneous thermal/physical effects during the first few thousand years of repository life. This is the time interval over which peak doses may occur from igneous activity, because beyond that period potential doses will diminish significantly due to the decay of shorter-lived radionuclides in the few waste packages involved in the volcanic activity (Mohanty et al., 2004). Even if the waste is directly exposed to magma because of package degradation after a long time period, quenching of the magma can produce a protective rind on the waste particles.

By not including the effects of magma solidification and quenching in the extrusive event scenario, important processes may not be adequately understood (e.g., those involved in entrainment and eruption of waste), and both the overall consequences and the risk of the package disruption process may be evaluated incorrectly. DOE's choice to use a conservative scenario to describe magma/waste package interactions does not justify overlooking insights gained by using a more realistic scenario.

Recommendation 1. Analysis of the consequences of an igneous dike intersection with a repository drift would be better risk informed by assessing the effects of magma solidifying upon entering a drift and quenching on the waste packages and any waste released from them. These studies could have an impact on conclusions regarding the number of waste packages that could be affected by a dike intrusion and the occurrence of secondary (flank) eruptions. This in turn would impact the amount of waste distributed in a resulting ash plume, the reasonably maximally exposed individual (RMEI) dose, and understanding of processes important to the total igneous activity effects.

EXPOSURE SCENARIO FROM CONTAMINATED EXTRUSIVE VOLCANIC MATERIALS

The NRC staff has updated the exposure scenario model incorporating particle size measurements from analogous volcanic eruptions. The revised and updated performance assessment model assumes a particle size distribution of dispersed contaminated ash with a median aerodynamic diameter of 10 microns and a minimum aerodynamic diameter of about 0.1 microns, thus including particulate matter that is not only inhalable but respirable.

The NRC staff's view, as presented to the Committee, is that long-term resuspension of contaminated fluvially dispersed ash and ash deposited on the surface can contribute to an inhalation dose to the RMEI. Consistent with this view, the NRC staff has selected parameter values for particle size distribution, dispersion, and long-term resuspension based on direct observation of volcanic ash at sites of recent volcanic activity. The Committee notes that these assumptions seem reasonable. Nonetheless, the Committee believes a more fully integrated analysis of the processes, parameters, and assumptions used in modeling this scenario would be helpful in making the staff's approaches transparent.

> Recommendation 2. The parameters and assumptions presented to date regarding the exposure scenario associated with igneous activity appear reasonable. However, in order to be adequately prepared for the license application review, the NRC staff should integrate all risk-significant aspects of the scenario by clearly justifying the processes, parameters and their values, and assumptions. The Committee believes the staff should use risk-informed approaches, including sensitivity studies, and other techniques to study and justify its choices.

PROBABILITY OF AN IGNEOUS DIKE INTERSECTING THE REPOSITORY

The NRC staff's single-valued estimate of the probability of an igneous intrusion, 10^{-7}/yr over the next 10,000 years, is at the higher end of the range of published estimates for dike intrusion on the order of 10^{-8}/yr to 10^{-7}/yr, authored by the NRC staff, their contractors, and the ACNW staff (Connor et al., 2000, Coleman et al., 2004).

> Recommendation 3. The NRC staff should reevaluate the use of a single value for probability of a volcanic intersection of the proposed Yucca Mountain repository, and should consider a range of estimates on the order of 10^{-7}/yr to 10^{-8}/yr based on studies published by NRC and previous ACNW views. If the staff decides to use a single-point value approach, the staff should document how this decision will support a risk-informed review of the consequences of an igneous event in a potential license application. Further evaluation of this range of probabilities should include consideration of new information being assembled for, and the results of, DOE's ongoing expert elicitation on Probabilistic Volcanic Hazard Assessment.

The Committee recognizes that some differences in views on volcanism between the ACNW and the NRC staff are a matter of professional judgment. The Committee appreciates the ongoing dialogue and offers its views as complementary to the staff's views. Consideration of these views may help the staff better risk inform their analyses of an igneous event during the potential Yucca Mountain license application review and related decisionmaking.

Work in progress by the NRC, which is unavailable to the Committee, may at least in part respond to the concerns addressed in this letter. Accordingly, the Committee plans to continue its dialogue with the NRC staff to better understand the bases of the staff's positions and to assess issues as additional information becomes available.

Sincerely,

Michael T. Ryan
Chairman

References:

1.   Coleman, N.M., B.D. Marsh, and L. Abramson. Testing Claims about Volcanic Disruption of a Potential Geologic Repository at Yucca Mountain, Nevada, Geophys. Res. Lett., doi:10.1029/2004GL021032, 2004.

2.   Connor, C.B., J. Stamatakos, D. Ferrill, B. Hill, G. Ofoegbu, F. Conway, B. Sagar, and J. Trapp. Geologic Factors Controlling Patterns of Small-volume Basaltic Volcanism: Application to a Volcanic Hazards Assessment at Yucca Mountain, NV, J. Geophys. Res., 105, 417–432, 2000.

3.   EPRI. Potential Igneous Processes Relevant to the Yucca Mountain Repository: Extrusive Release Scenario: Analysis and Implications. EPRI Report 1008169. Electric Power Research Institute: Palo Alto. 2004.

4.   Mohanty, S., et al. System-level Performance Assessment of the Proposed Repository at Yucca Mountain Using the Tpa Version 4.1 Code. NRC accession no. ML041350316. 2004.

UNITED STATES
NUCLEAR REGULATORY COMMISSION
ADVISORY COMMITTEE ON NUCLEAR WASTE
WASHINGTON, D.C. 20555-0001

December 9, 2005

The Honorable Nils J. Diaz
Chairman
U.S. Nuclear Regulatory Commission
Washington, D.C. 20555-0001

SUBJECT: DEVELOPMENT OF A STANDARD REVIEW PLAN FOR U.S. DEPARTMENT
OF ENERGY WASTE DETERMINATIONS

Dear Chairman Diaz:

The U.S. Department of Energy (DOE) is expected to pursue a number of determinations that
certain wastes are not high-level waste as a prerequisite to allowing disposal. DOE is required
or expected to request that the U.S. Nuclear Regulatory Commission (NRC) perform technical
reviews of the Department's waste determinations and, in some cases, its disposal and
monitoring plans for the wastes.[1] The NRC staff is currently developing a Standard Review
Plan (SRP) for these reviews. In this letter the Advisory Committee on Nuclear Waste provides
its recommendations on the development of the SRP based on information obtained from the
following activities:

- The Committee held a 2-day public working group meeting on waste determination
  August 2 - 3, 2005, during its 162nd meeting. The working group meeting included
  background presentations by DOE and NRC staff; 12 presentations by experts from
  academia, research institutions, and private enterprise; three panel discussions
  involving these same experts and staff from the NRC Office of Nuclear Regulatory
  Research; and input from State agencies and public stakeholders.

- Three Committee members, ACNW staff, the Director of the Division of Waste
  Management and Environmental Protection in the Office of Nuclear Material Safety and
  Safeguards (NMSS), and a member of the public made a 1-day visit to the Savannah
  River Site (SRS) on August 10, 2005. They toured the tank farms, tank waste
  processing facilities, waste vitrification facilities, and equipment development facilities.
  The participants also benefitted from formal and informal discussions with SRS staff
  about their approach to tank cleanup and waste determinations.

- Members of the Committee, ACNW staff, and NRC staff toured the West Valley
  Demonstration Project (WVDP) site, participated in a working group meeting, and heard
  input from the public on October 18 - 20, 2005.

---

[1] Section 3116 of the Ronald Reagan National Defense Authorization Act (NDAA) of
Fiscal Year 2005 (Public Law 108-375-October 28, 2004) makes the NRC responsible for
providing technical consultation to DOE on waste determinations in the States of South
Carolina and Idaho and, in coordination with the concerned State, for monitoring DOE disposal
actions.

- A Committee member who is also a member of a National Research Council committee addressing issues related to waste determinations visited DOE's tank waste storage sites at SRS, Hanford, and Idaho National Laboratory.

- An ACNW staff member attended a demonstration of waste retrieval technologies in Mooresville, North Carolina on September 7, 2005.

- An ACNW staff member attended a briefing to the National Research Council's Nuclear & Radiation Studies Board on previous and ongoing studies of issues related to waste determinations held in Washington, D.C., on September 12, 2005.

Based on the information obtained from these activities, the Committee developed the observations and recommendations provided in this letter. The observations and recommendations are organized as follows:

- Section 1 concerns the overall scope of the SRP.

- Section 2 addresses the overall consistency among criteria for waste determinations as well as the consistency of performance objectives and key phrases in the criteria, and the consistency of the criteria with other NRC regulations and guidance.

- Section 3 provides insights concerning evaluation of two key components of waste determinations: the status of radionuclide removal technology and performance assessment.

- Section 4 addresses how to evaluate whether wastes have been removed to the "maximum extent practical" and whether doses are "as low as reasonably achievable (ALARA)."

- Section 5 addresses technical considerations regarding NRC guidance on monitoring of waste determined to not be high-level waste to assess compliance with the performance objectives of Subpart C to 10 CFR Part 61.

1. STANDARD REVIEW PLAN SCOPE

The principal purposes of an SRP are to enhance the quality and uniformity of staff reviews and to present a well-defined base from which to evaluate proposed changes in the scope and requirements of reviews. The NRC has experience in developing and implementing SRPs in program areas related to waste determination reviews. The most relevant technical information can be found in NUREG-1200, "Standard Review Plan for the Review of a License Application for a Low-Level Radioactive Waste Disposal Facility," Revision 3, April 1994; NUREG-1757, "Consolidated NMSS Decommissioning Guidance," September 2003 along with draft Supplement 1, issued for public comment in September 2005; and NUREG-1573, "A Performance Assessment Methodology for Low-Level Radioactive Waste Disposal Facilities: Recommendations of NRC's Performance Assessment Working Group," October 2000.

Guidance on risk-informed, performance-based approaches helpful to the development of the SRP can be found in NUREG-1549, "Decision Methods for Dose Assessment to Comply with Radiological Criteria for License Termination-Draft Report for Comment," July 1998, and the June 1998 SRM-SECY-98-144 on the staff's white paper on risk-informed and performance-based regulation.

Developing the SRP for waste determinations is complicated by the diversity of radioactive materials to be considered, the existence of multiple sets of criteria for developing and reviewing waste determinations, and the NRC's role as consultant instead of statutory regulator.

Recommendation: The SRP should be a single document that provides integrated guidance to NRC staff on risk-informed reviews of waste determinations and implicit guidance to DOE on the information to be provided in the waste determination. The waste determination SRP should build on the generic format, content, and implementation of existing SRPs and on relevant information in existing SRPs. The Committee believes the integration will enhance uniformity and efficiency of the reviews.

2.   CONSISTENCY

2.1   Criteria for Determinating of Waste Classification

The criteria for preparing and reviewing a waste determination depend on the specific waste and site:

*   Section 3116 of the NDAA is applicable to some waste determinations at Savannah River and Idaho.

*   NRC Decommissioning Criteria for the West Valley Demonstration Project (M-32) at the West Valley Site, Final Policy Statement [64 FR 67952, December 3, 1999] are applicable to some waste determinations there.

*   DOE Order 435.1, "Radioactive Waste Management," and the supporting documents DOE M 435.1-1, "Radioactive Waste Management Manual," and DOE Guide 435.1-1, "Implementation Guide for use with DOE Manual 435.1-1," issued in 1999 and reissued in 2001, may be used as a basis for some waste determinations.

Recommendation: The SRP should adopt a consistent technical interpretation of similar criteria in the three sets of criteria.

2.2   Subpart C Performance Objectives

The Committee notes that under Section 3116 NRC staff must review waste determinations to assess conformance with 10 CFR Part 61 Subpart C performance objectives. The other two sets of criteria allow disposal to meet safety objectives comparable to the objectives stated in Subpart C. The Committee believes that the SRP should focus on confirming that DOE's proposed safety objectives are essentially identical to those in Subpart C.

Recommendation: The SRP should accept use of Subpart C performance objectives per se in all sets of criteria. If DOE chooses to use a different set of objectives, the SRP should expect DOE to provide a compelling technical justification to show that the objectives are as protective as those in Subpart C.

2.3 "Highly Radioactive" and "Key" Radionuclides

DOE Manual 435.1 and the WVDP criteria use the phrase "key radionuclides" in addressing radionuclide removal, whereas the Section 3116 criteria use the phrase "highly radioactive radionuclides." "Highly radioactive" commonly refers to relatively short-lived radionuclides, particularly if they emit penetrating radiation. The Committee notes that this common interpretation would not lead to a risk-informed approach because (a) it excludes long-lived radionuclides that should be removed to the maximum extent practical because they are important to risk in many situations (e.g., Tc-99, Np-237) and (b) it is based on an inherent property of a radionuclide (its decay characteristics) instead of the risk posed by the waste of which the radionuclide is a part. The Committee believes a risk-informed interpretation of "highly radioactive" and "key" radionuclides can best be accomplished by analyzing the results of a risk-informed performance assessment for the radionuclides that are the dominant contributors to dose.

Recommendation: The SRP should adopt a risk-informed interpretation of "highly radioactive radionuclides" by defining it to mean the same as "key radionuclides," i.e., radionuclides potentially important to meeting the Subpart C performance objectives.

2.4 Other NRC Regulations and Guidance

After removal, processing, and conversion to a solid form, tank waste will be disposed of in much the same way as is waste at a commercial low-level waste (LLW) disposal facility. A grout and cap approach is typically planned for in-place isolation of residual waste in tanks. This approach has similarities to site decommissioning. Existing NRC regulations and guidance in these two areas reflect years of experience. Examples of such guidance documents for performance assessments are NUREG-1573, NUREG-1757, and NRC staff "Technical Position on Waste Form", Revision 1, January 1991.

Recommendation: Existing NRC regulations and guidance documents should be used as a source of insights for developing the SRP.

3. TECHNOLOGY AND PERFORMANCE ASSESSMENT

3.1 Technology for Removal of Radionuclides

The Committee notes that DOE has many waste retrieval and radionuclide separation technologies available and has been relatively successful in its completed retrieval efforts. However, the Committee believes DOE will continue to face technical challenges in radionuclide removal because it has retrieved only a small portion of the waste that will eventually require retrieval and most of this waste has been retrieved from DOE's less complex tanks. Furthermore, DOE has separated radionuclides from only a fraction of the retrieved waste. The Committee observes that DOE continues to improve its radionuclide removal technologies and adopt new technologies to address challenges as they arise.

Recommendation: The NRC staff should review the approaches to waste retrieval and radionuclide separation in each waste determination in the context of relevant existing and projected technologies. The staff should expect DOE to have considered existing relevant technologies or technologies being developed by domestic and international organizations.

3.2   Performance Assessment

Historically, variability and uncertainty in performance assessments for near-surface waste disposal were addressed by selecting one or two different values for parameters believed to be important and observing how much the estimated dose from a deterministic performance assessment changes. Exclusion of probabilistic performance assessments has been justified by using conservative approaches in the deterministic performance assessment.

The Committee believes that assumptions such as the duration of effective institutional controls and selection of conceptual models such as those for groundwater flow can dominate the magnitude of the estimated dose from near-surface waste disposal facilities. Many assumptions such as those about institutional control cannot be validated because they involve predictions of the future behavior of people and there is a growing body of literature citing experience which raises concerns about the reliability of such controls. Conceptual models of physical systems are theoretically amenable to validation through analysis or testing, but many situations are so complex that validation may not be practical.

The Committee notes the extensive use of cementitious materials as structural barriers and solid matrixes for isolating, in near-surface disposal facilities, wastes determined to not be high-level waste. Assumptions about the rates at which the beneficial properties of cementitious materials degrade are therefore important to the results of performance assessments for such facilities.

Recommendation: The SRP should specify a preference for probabilistic performance assessments using best estimates with explicit analysis of uncertainties. Exceptions should include documentation of how uncertainties were addressed.

Recommendation: The SRP should recognize that some important performance assessment assumptions are incapable of validation. Such assumptions should be based on realistic consideration of empirical evidence to the extent such evidence exists and should be subjected to uncertainty analyses.

Recommendation: The SRP should provide guidance to the NRC staff on reviewing improvements in technical bases for assumptions concerning the long-term degradation rate of cementitious materials in waste disposal applications. The NRC staff should maintain the capability to review justifications for performance assessment assumptions based on cutting-edge research concerning cementitious materials.

4. "MAXIMUM EXTENT PRACTICAL" AND ALARA

All three sets of criteria require that the amount of radionuclides in a waste be reduced to the "maximum extent practical" or the "maximum extent technically and economically practical," and that doses to workers or the public be ALARA. All of these goals are functionally the same: they require that factors such as the capability of technologies, costs, and risks associated with competing radionuclide removal alternatives be evaluated as a basis for deciding how much risk reduction (i.e., waste retrieval and processing, and use of engineered barriers) is enough. The potential importance of risks posed by other nearby waste disposal areas and contaminated environmental media is a factor to be considered in making this decision.

The Committee observes that complex decisions are likely to require consideration of stakeholder values and demands as well as technical issues. The waste determination decisionmaking process and the process for developing the SRP should be transparent and allow stakeholder participation. The November 10, 2005, NMSS public scoping meeting[2] to obtain input on the development of the SRP, was a good start toward achieving this goal.

Recommendation: The information necessary to support DOE's determination that radionuclides have been removed to the maximum extent practical or maximum extent technically and economically practical, and that estimated doses are ALARA should be the same for all sets of criteria.

Recommendation: A risk-informed evaluation of ALARA or radionuclide removal to the maximum extent practical or maximum extent technically and economically practical should be done in the context of the surrounding risk.

5. MONITORING TO ASSESS COMPLIANCE WITH PERFORMANCE OBJECTIVES OF SUBPART C

Under provisions of Section 3116, the NRC, in coordination with the host State, is required to monitor DOE disposal actions for the purpose of assessing compliance with the Subpart C performance objectives. The Committee believes compliance monitoring should be considered in the design of a system to isolate waste and the associated performance assessment. The Committee further believes that the types and quantities of waste likely to be disposed of onsite should be considered in selecting monitoring approaches and systems.

Recommendation: NRC staff activities to determine compliance with Subpart C performance objectives should review the design of barriers to radionuclide release to ensure that provisions have been made for future monitoring activities. Engineered barrier design has already been completed for some waste determinations. For these cases, the NRC will have to rely on reviewing the adequacy of the designs and determining whether improvements are necessary or feasible.

---

[2] Attended by Committee Vice-Chairman Allen Croff and Committee staff member Latif Hamdan

Recommendation: Far-field and near-field monitoring, engineered barrier monitoring, and performance assessment are key elements of performance confirmation. The SRP should provide guidance to the NRC staff on these topics that includes information on how waste disposal facilities can be designed to facilitate monitoring.

The Committee looks forward to reviewing the draft SRP as the document evolves. As a result of the future opportunities for the Committee to provide its input, it does not expect a formal response to this letter from NRC staff in favor of allowing them to focus their energies on preparing the draft SRP.

Sincerely,

Michael T. Ryan
Chairman

December 23, 2005

The Honorable Nils J. Diaz
Chairman
U.S. Nuclear Regulatory Commission
Washington, D.C. 20555-0001

SUBJECT: WEST VALLEY DEMONSTRATION PROJECT (WVDP) - ACNW WORKING
GROUP MEETING

Dear Chairman Diaz:

The ACNW held a working group meeting on the WVDP on October 19, 2005. The meeting was held in Ellicottville, NY, a location close to the site. Three experts[1] on performance assessment and the decommissioning of complex sites participated at the invitation of the Committee. Members of the Committee and two of the invited experts toured the site on October 18. The meeting was attended by representatives of several New York State agencies, including the New York State Energy Research and Development Authority and the New York State Departments of Health and Environmental Conservation, the NRC staff, stakeholders, and the general public.

The purpose of the meeting was to receive an update on the status of decommissioning activities at the site and to hear about DOE's and NRC's approaches to the WVDP site performance assessment. The performance assessment work is still in the early stages and many decommissioning activities are ongoing.

The Committee has the following observations:

- The WVDP is an important case study that provides a useful model for the decommissioning of complex sites. Decommissioning activities at WVDP will need to address spent fuel, disposition of vitrified high-level waste, drummed and grouted supernatant, tanks containing very high levels of residual radioactive material, various buildings, NRC and NY State-licensed landfills, and soil and groundwater contamination.

- The decommissioning is especially complicated because the site is owned by two parties and because several Federal and State agencies are responsible for the remedial activities.

---

[1] The invited experts were Dr. David C. Kocher, Senior Scientist, SENES Oak Ridge, Inc. (a consultant to the Committee); Dr. Frank Parker, Distinguished Professor of Environmental and Water Resources Engineering, Vanderbilt University; and Mr. Thomas Nauman, Vice President and Northeast Regional Director, Shaw, Stone & Webster, Inc.

- Ongoing erosion adjacent to the landfills raises concerns about potential exposure of buried waste. Erosion modeling and analysis will be critical to decisionmaking concerning long-term protection.

- DOE and NRC are taking different approaches to performance assessment for the WVDP. Whereas DOE's approach is primarily deterministic, the NRC staff will use a probabilistic approach that will enable a risk-informed review. The Committee believes that the staff approach to performance assessment is technically sound and commends the staff for taking a risk-informed approach.

- Further characterization of the subsurface strontium plume is needed to provide a better basis for decisionmaking. Existing strontium data can be used to verify the groundwater modeling and build confidence in the modeling predictions.

The Committee offers the following recommendations:

- The erosion analysis will be critical to the choices of remedial technology for the landfills. The Committee recommends that the staff be well prepared to review the DOE erosion modeling.

- Subsurface characterization data, including groundwater monitoring data for strontium, while not yet sufficient for remedial technology evaluation, can be used to verify groundwater modeling and build confidence in the modeling predictions. The Committee encourages the staff to use these data in its performance assessment.

The Committee looks forward to hearing the results of the performance assessments and the associated environmental impact statement and to having further interactions with the staff on this very complex site evaluation.

Sincerely,

Michael T. Ryan
Chairman

December 27, 2005

The Honorable Nils J. Diaz
Chairman
U.S. Nuclear Regulatory Commission
Washington, D.C. 20555-0001

SUBJECT: OBSERVATIONS OF STAKEHOLDER PARTICIPATION IN RECENT
             MEETINGS OF THE ADVISORY COMMITTEE ON NUCLEAR WASTE (ACNW)

Dear Chairman Diaz:

Part of the ACNW's action plan is to encourage public participation as a way to enhance openness and make the deliberations and decisions of the agency more transparent. During the last 6 months, the Committee has arranged and conducted meetings in locations other than NRC headquarters to encourage public participation and listen to stakeholders' concerns and perspectives. Meetings were held at locations and times to facilitate stakeholder participation. In addition, the Committee participated in public meetings held by the NRC Office of Nuclear Materials Safety and Safeguards (NMSS) to observe and hear the views of stakeholders and members of the public. Finally, individual members attended public meetings relevant to the Committee's action plan.

OBSERVATIONS

From recent ACNW meetings in New York, Nevada, South Carolina, and Maryland, the Committee provides the following observations.

1.   The views that the ACNW heard during its public meetings assisted the Committee in its deliberations by providing additional perspectives and insights.

2.   NMSS held a public workshop on its proposed revisions to decommissioning guidance. The Committee believes that participating in such public workshops is an effective way for the Committee to gain insights from groups of stakeholders and members of the public.

3.   The Committee derives significant value from attending public meetings pertinent to its action plan.

DISCUSSION

(1) West Valley, New York

The Committee reviewed work being done to support the decommissioning process for the West Valley Demonstration Project (WVDP). The Committee believes this site is likely to employ all of the options outlined in the Commission's License Termination Rule. On

October 18-20, 2005, the Committee toured the WVDP site and held a public meeting in Ellicottville, New York (about 10 miles from the WVDP). The meeting consisted of briefings on the NRC's and Department of Energy's (DOE's) performance assessment methodologies and some insights derived from them. DOE described decommissioning remediation work underway at the site. Representatives of the West Valley Citizen's Task Force, Canbury Agency, the Coalition on West Valley Nuclear Waste, and the Nuclear Information and Resource Service also offered comments to the Committee.

Approximately 75 people attended the meeting, including members of the community and representatives of the New York State Energy Research and Development Authority (NYSERDA), the New York State Department of the Environmental Conservation, and the New York State Department of Health. Views and concerns were expressed about the movement of radioactive materials in groundwater and potential erosion leading to exposure of buried waste. New York State agency representatives gave their perspectives on and described their roles for the future of the West Valley Site. A representative of Clark County, Nevada, was also in attendance. The transcript of the meeting contains these comments and documents a range of views that were helpful to the Committee in its deliberations.

(2) Las Vegas, Nevada

The agency's strategic plan identifies hosting public meetings in Nevada and along major transportation corridors to the proposed Yucca Mountain repository as a means to support its openness strategy. The ACNW typically holds meetings once a year in Nevada to take advantage of the availability of onsite technical personnel working on the Yucca Mountain project and to solicit stakeholders' concerns and views. Committee members and staff toured the Yucca Mountain site. Two stakeholder representatives participated in the site visit. The Committee conducted its public meeting on September 21, 2005, at the Atomic Safety and Licensing Board Panel's (ASLBP's) new facility for future hearings on the repository. The meeting was not well attended by members of the public, perhaps because the ASLBP facility is new and is located well away from the usual venues for public meetings in Las Vegas. Some members of the public said there would have been more participation if the meeting were closer to the Yucca Mountain area.

Representatives from the Las Vegas Paiute tribe, the State of Nevada's Nuclear Waste Project Office, Clark County (Nevada), the Nevada Nuclear Waste Task Force, the Los Angeles Environmental Technology Center, and Monitor Scientific made comments at the meeting. Mr. Mike Henderson of Congressman Jim Gibbon's office provided a written statement and read it into the record of the meeting. In addition, unaffiliated members of the public provided comments. The comments were generally not supportive of activities at the proposed Yucca Mountain project. Commenters expressed concerns about emergency evacuation plans, proposed revisions to the Environmental Protection Agency's standards, repository performance, and the impact on citizens along transportation routes in Nevada. The transcript contains the record of these comments.

(3) DOE Savannah River Site and the Barnwell Low-Level Waste Disposal Facility

On August 11, 2005, the Committee toured and held discussions at the Barnwell low-level radioactive waste (LLW) facility. The purpose of the visit was to familiarize the Committee members and ACNW staff with current low-level waste disposal technologies and practices and to obtain information on low-level waste activities that should be reflected in the Committee's FY 2006 Action Plan. The Barnwell facility will cease to accept LLW from generators outside the Atlantic Compact in 2008.

Committee members also discussed Barnwell site issues during a lunch that was attended by approximately 20 participants. Elected representatives and officials from Barnwell County and City said there was significant and strong support in the community for the LLW facility. Representatives of Chem-Nuclear/Duratek, the contract manager for the site, accompanied and assisted the Committee during its visit. An inspector from the State of South Carolina Department of Health and Environmental Control accompanied the members throughout the site visit. Other participants included representatives of Barnwell County Council, the Barnwell County Chamber of Commerce, and the Barnwell County Economic Development Commission. Both the Chem-Nuclear/Duratek representatives and the State inspector said the regulator and the licensee communicate "openly and honestly" on a routine basis.

The community leaders with whom the Committee members spoke are in favor of expanding the role of the facility to receive LLW. They also indicated that the Chem-Nuclear/Duratek management and staff have been open and honest with the community and are good corporate citizens.

Barnwell's decommissioning and long-term stewardship funds were discussed. The decommissioning fund is held by a third-party trustee. The stewardship fund covers long-term institutional controls and is deposited with the State. The decommissioning fund held by a third party is intact. The long-term stewardship fund was intact a few years ago at $104M. However, since 2002 the State of South Carolina has borrowed $90M from this fund. The current administration has promised that $25M will be restored this fall.

On August 10, 2005, the Committee visited the Department of Energy's Savannah River Site (SRS) to obtain insights on waste operations and processes. The insights will help the members prepare advice on the NRC standard review plan for waste determinations. The SRS personnel briefed the members on tank waste management, waste retrieval technology, sludge washing, disposal of grouted waste in near-surface vaults, tank closure, and radionuclide separation. The Committee was also briefed on plans for the proposed mixed-oxide fuel fabrication facility.

A representative of Clark County, Nevada, participated in both South Carolina site visits.

(4) Participation in the NMSS Workshop on Proposed Revisions to Decommissioning Guidance

The Committee participated in the NRC staff's April 2005 workshop in Gaithersburg, Maryland. The objectives of the workshop were to:

- Inform stakeholders of NRC's Integrated Decommissioning Improvement Plan (IDIP), including regulatory and program management improvements.

- Discuss the NRC staff's development of proposed revisions to guidance for decommissioning under the License Termination Rule.

- Solicit feedback and suggestions from stakeholders on proposed revisions to decommissioning guidance and on decommissioning lessons learned.

Approximately 180 members of the public and stakeholders attended the workshop. The information obtained in this meeting was useful to the Committee in its deliberations.

(5) Other Meetings Attended by Committee Members

Pursuant to the Committee's action plan, members attended various public meetings as observers:

- The Environmental Protection Agency's meeting on the proposed revised standard for Yucca Mountain on October 4 and 5, 2005, in Las Vegas, Nevada

- The LLW Forum, Inc., meeting on radioactive waste on September 23, 2005, in Las Vegas, Nevada

- The NMSS public scoping meeting on the waste determination standard review plan on November 9, 2005, in Gaithersburg, Maryland

- The interagency meeting on research concerning the performance of cementitious materials on June 2, 2005 in Rockville, Maryland

- The DOE Workshops on Probabilistic Volcanic Hazard Analysis on February and August, 2005 in Las Vegas, Nevada

The Committee will continue to seek and encourage broader public participation in its CY 2006 meetings, including topical working group meetings. The Committee will also observe stakeholder meetings organized by NRC and other Federal agencies concerning activities within the scope of the Committee's action plan.

Sincerely,

Michael T. Ryan
Chairman

UNITED STATES
NUCLEAR REGULATORY COMMISSION
ADVISORY COMMITTEE ON NUCLEAR WASTE
WASHINGTON, D.C. 20555-0001

December 27, 2005

The Honorable Nils J. Diaz
Chairman
U.S. Nuclear Regulatory Commission
Washington, D.C. 20555-0001

SUBJECT: OPPORTUNITIES IN THE AREA OF LOW-LEVEL RADIOACTIVE WASTE
MANAGEMENT

Dear Chairman Diaz:

At a briefing of the Commission earlier this year, the Advisory Committee on Nuclear Waste
(ACNW) offered to review the history of the U.S. Nuclear Regulatory Commission's (NRC's)
low-level radioactive waste (LLW) regulatory framework and identify areas for which the
framework might be better risk-informed. The Committee and its staff developed the enclosed
white paper. This white paper provides a thorough, though not exhaustive, examination of the
history and status of NRC's LLW regulatory program, based on a review of the available
literature. The paper includes a summary of past ACNW advice to the Commission in this area.

The Committee believes that current regulations are fully protective of the public health and
safety and fully protective of worker health and safety. The Committee also believes that this
white paper provides a framework to identify opportunities to better risk-inform and improve the
effectiveness of LLW management and regulation. The Committee believes the white paper
will contribute to future work with staff and stakeholders. In its FY 2006-2007 Action Plan, the
Committee recommends working group meetings to address specific LLW activities. The
Committee also believes that where possible the improvements in risk-informing LLW
regulations should be accomplished through licensing actions and regulatory guidance.

At its 165th meeting on December 13, 2005, representatives of the Office of Nuclear Materials
Safety and Safeguards (NMSS) briefed the Committee on current staff activities in the area of
LLW and on their preliminary views on the Committee's white paper. Development of this white
paper provided a vehicle for interacting constructively with the NRC staff. The staff provided
comments which have helped to improve the technical content and format of the paper. During
the last two Committee meetings, LLW stakeholders had an opportunity to recommend
information for inclusion in the paper to ensure the completeness of the history of the LLW
regulatory program. The white paper also includes a list of stakeholders that have published
recent positions on LLW management issues.

The Committee has learned that the staff is initiating a strategic planning effort to identify and
prioritize the agency's major LLW activities based on the issues that have emerged over the
last few years. The staff plans to select activities that will lead to improvements in the current
U.S. LLW disposal system.

Some of the activities being considered include, but are not limited to:

- Reviewing past guidance on LLW storage
- Responding to a 2005 Commission order on the disposal of large quantities of depleted uranium
- Addressing 10 CFR 20.2002 issues

There are other related external efforts and initiatives underway, including the National Academy of Sciences low-activity waste study (Phase II) and a new Government Accountability Office review of best LLW management practices. The Committee is aware of these activities, will continue to follow them, and will consider the findings when they become available.

At this time, the Committee believes that it is prudent to identify a preliminary list of areas where Part 61 might be better risk-informed. The list can be expanded later, depending on the outcomes of the future efforts and reviews. The Committee believes it must communicate with staff to better leverage the Committee's work to support the staff's strategic planning in the LLW area. The Committee also believes it is important to include stakeholder views in identifying opportunities to improve the effectiveness of the regulatory framework. It is also important to identify and evaluate any unintended consequences from recommended changes. Part 61 is referenced in a number of other regulations and laws, and any opportunities for risk-informed improvements need to be carefully assessed for their impacts in these other areas.

The Committee does not intend the following list of opportunities for risk-informing LLW regulation to be exhaustive or to reflect any ranking or priority.

- Part 61 intruder scenarios are not risk-informed. They are based on bounding or extremely conservative assumptions and conditions. Furthermore, there is no guidance on performing an LLW human intrusion calculation. The assumptions used in the intruder scenario have a direct bearing on the Class A, B, and C concentration limits in Section 61.55. Section 61.58 allows for alternative requirements for waste classification and characteristics. This section could serve as a basis for better risk-informing 10 CFR 61.55.

- Part 20 has been updated to incorporate recent recommendations of the International Commission on Radiological Protection (ICRP) Section 61.41 relies on older ICRP dosimetry models that are based on a different system of dose calculation. This inconsistency can cause confusion.

- With one exception, the Subpart D siting criteria are qualitative. A more quantitative and risk-informed or performance-based approach to siting criteria might be helpful in developing new sites.

- The Part 61 institutional controls and financial assurance measures have recently been considered in the proposed revision to decommissioning guidance. The updates may provide insights into the institutional control and financial assurance requirements for LLW sites.

- Collection of environmental monitoring data is required during the operational and institutional control periods. These data could be used to increase confidence in long-term predictions of performance of LLW facilities.

- Credit for engineered barriers for waste form, waste packaging, disposal site design, and cover design were not explicitly included in Part 61. It would be an improvement to consider appropriate credit for the contribution of these engineered features to system performance.

We are forwarding the LLW White Paper as a draft final version, subject to limited peer review. We plan to issue the final version shortly as a NUREG report. Using the white paper as a starting point, the Committee is prepared to interact with the NMSS staff and stakeholders on risk-informing the management of LLW. Because of significant stakeholder interest in LLW activities, the ACNW plans to sponsor a working group meeting with NMSS to solicit stakeholder views on what changes to the regulatory framework for managing LLW should be recommended for Commission consideration.

Sincerely,

Michael T. Ryan
Chairman

Enclosure: ACNW White Paper

ACNW WHITE PAPER:
HISTORY AND FRAMEWORK OF COMMERCIAL LOW-LEVEL RADIOACTIVE WASTE MANAGEMENT IN THE U.S.

Submitted by

The Advisory Committee on Nuclear Waste

December 30, 2005

CONTENTS

APPENDICES

TABLES

ABBREVIATIONS

ACNW	Advisory Committee on Nuclear Waste
AEA	Atomic Energy Act of 1954
AEC	Atomic Energy Commission
AIF	assured isolation facility
ALARA	as low as reasonably achievable
ANPR	Advance Notice of Proposed Rulemaking
BEIR	biological effects of ionizing radiation
BRC	below regulatory concern
BWR	boiling water reactor
CAA	Clean Air Act of 1977, as amended
CEQ	Council on Environmental Quality
CFR	*Code of Federal Regulations*
CORAR	Council on Radionuclides and Radiopharmaceuticals
DCFs	disposal concentration guides
DSI	Direction-Setting Initiative
DOE	U.S. Department of Energy
DOT	U.S. Department of Transportation
EMCB	earth-mounded concrete bunker
EIS	environmental impact statement
EPA	U.S. Environmental Protection Agency
ERDA	Energy Research and Development Administration
FR	*Federal Register*
GAO	U.S. General Accounting Office
GTCC	greater-than-Class C radioactive waste
HAPS	hazardous air pollutants
HIC	high-integrity container
HLW	high-level radioactive wastes
HPDE	high-density polyethylene
ICRP	International Commission on Radiation Protection
ISFSI	independent spent fuel storage installation
LLW	low-level radioactive waste
LLW Forum	Low-Level Radioactive Waste Forum
LLWPA	Low-Level Waste Policy Act of 1980
LLWPAA	Low-Level Waste Policy Amendments Act of 1985
LSV	liquid scintillation vial
LWR	light-water reactor
MAC	maximum average concentration
MCLs	maximum concentration limits
MIMS	Manifest Information Management System (of DOE)
MOU	Memorandum of Understanding
MRS	monitored retrievable storage
NACOA	National Advisory Committee on Oceans and the Atmosphere
NARM	naturally-occurring and accelerator-produced radioactive material
NAS	National Academy of Sciences
NBS	National Bureau of Standards
NCRP	National Council on Radiation Protection and Measurements
NEPA	National Environmental Policy Act of 1970

NESHAPS	National Emission Standards for Hazardous Air Pollutants
NOI	Notice of Inquiry
NORM	naturally occurring radioactive material
NRC	U.S. Nuclear Regulatory Commission
NUMARC	Nuclear Management and Resources Council
OMB	Office of Management and Budget
ORNL	Oak Ridge National Laboratory
OTA	Office of Technology Assessment
PAM	performance assessment methodology (for LLW)
PCBs	polychlorinated biphenyls
PRA	probabilistic risk assessment
PRESTO	Protection of Radiation Effects from Shallow Trench Operations (EPA computer code)
PWR	pressurized water reactor
QA	quality assurance
RCRA	Resource Conservation and Recovery Act of 1976
RES	Office of Nuclear Regulatory Research
SA	specific activity
SLB	shallow land burial
SNF	spent nuclear fuel
SNL	Sandia National Laboratory
SRM	Staff Requirements Memorandum
SS	sources and special nuclear material
TEDE	total effective dose equivalent
TENORM	technologically-enhanced naturally occurring radioactive materials
TRU	transuranic radioactive waste
TSCA	Toxic Substances Control Act of 1976
USGS	U.S. Geological Survey
WCS	Waste Control Specialists, LLC (of Texas)

PART I – LOW-LEVEL RADIOACTIVE WASTE PROGRAM HISTORY

1   Early Approaches to the Management of Low-Level Radioactive Waste

Most establishments working with radioactive materials produces radioactive wastes since anything the radioactive material comes into contact with becomes contaminated. In the United States, thousands of establishments, both government and private, are licensed to use radioactive materials. The volume and level of activity in the wastes produced varies in direct proportion to the amount of radioactive material used. Historically, in the United States, the greatest proportion of radioactive waste produced is low-level radioactive waste (or LLW); although LLW only accounts for about 0.1 percent of the total radioactivity being disposed of. See Moeller (1992, p. 118).

The term "low-level radioactive waste" or "LLW" has carried a changing and somewhat imprecise definition over the years. Before the promulgation of the U.S. Nuclear Regulatory Commission's (NRC or the Commission) LLW disposal regulations found at Title 10, Part 61, "Licensing Requirements for Land Disposal of Radioactive Waste," of the *Code of Federal Regulations* (10 CFR Part 61), the term LLW was exclusionary. It generally meant that portion of the radioactive waste stream that did not fit the prevailing definition of high-level (HLW) or intermediate-level radioactive wastes at the time, and with concentrations of transuranic elements less than 100 nanocurries per gram (nCi/gm). Some LLW has radioactive material concentrations comparable to that of spent nuclear fuel (SNF) and this waste is considered by the NRC to be greater-than-Class C (GTCC) radioactive waste. Such wastes are the responsibility of the U.S. Department of Energy (DOE) to manage. Some LLW contains chemically hazardous constituents and is referred to as "mixed waste." Some of this type of LLW is subject to regulation under the provisions of the Resource Conservation Recovery Act of 1976 (RCRA).[1]

Commercial sources generate less than 34 percent of LLW by volume. It is generated at commercial nuclear power plants, research laboratories, hospitals, industrial facilities, and universities, most of whom are NRC or Agreement State licensees. Some LLW is generated in facilities that are not regulated under the NRC's authority under the Atomic Energy Act (AEA) but are regulated by the States. LLW typically consists of contaminated protective shoe covers and clothing, wiping rags, mops, filters, reactor water treatment residues, equipments and tools, luminous dials, swabs, injection needles, syringes, and laboratory animal carcasses and tissues. The radioactive material concentration can range from just above background levels found in nature to very high concentrations of radioactive material in certain cases such as parts from inside the reactor vessel in a nuclear power plant. About 97 percent of LLW decays to safe levels within 100 years whereas a small percentage of longer-lived radionuclides persist at potentially hazardous concentrations through 300 to 500 years. LLW is typically stored on-site by licensees, either until it has decayed and can be disposed as ordinary trash, or until

---

[1] RCRA is administered by Environmental Protection Agency (EPA) as well as States with comparable RCRA regulations. RCRA defines a hazardous waste as any substance that is flammable, corrosive, reactive, or toxic. RCRA-classified waste must be managed and disposed in compliance with regulations for both chemical and radiological hazards. Due to legislative ambiguity (Parler, 1989), the management of mixed waste is subject to dual regulation by both NRC and the EPA. However, the review of mixed waste issues is beyond the scope of this paper. The management of mixed wastes is discussed at some length in Office of Technology Assessment (OTA – 1989) and the National Academy of Sciences (NAS – 1999).

amounts are large enough for shipment to an approved disposal site. DOE, operating under different rules from the commercial sector, disposes of much of its own LLW.[2] Government-generated LLW includes past nuclear weapons production and research, environmental restoration of federal facilities, and routine operations of the U.S. Navy's nuclear propulsion program. Review of DOE's LLW management program is beyond the scope of this paper but is described in a National Safety Council (2002) report.

1.1   Ocean Disposal

In the early years of the domestic atomic energy industry, the Atomic Energy Commission (AEC)[3] used three methods to dispose of radioactive waste – dilution and dispersion, shallow land burial (SLB), and disposal at sea.

Both commercial and noncommercially-generated LLW was first disposed only by the AEC of 1954. Commercial wastes were typically disposed in the ocean, based on the recommendations of the NAS (1959, 1962).[4] Because most radionuclides had short half-lives, it was believed that dilution in ocean water plus decay would result in innocuous levels and pose minimal hazards to man. Furthermore, there was the view that the sea was readily available and economic to use (Raubvogel, 1982; pp. 21–23). Disposal at sea was conducted by the U.S. Navy until about 1959. Thereafter, the AEC licensed six companies to dispose of the wastes.

Ocean disposal of LLW occurred in waters greater than 1000 fathoms (about 6000 ft) following the 1954 recommendations of the National Bureau of Standards' (NBS) *National Committee on Radiation Protection* (NBS, 1954; p. 2). The disposal container most often used was a 55-gallon steel drum. The LLW was mixed with cement or concrete to assure sinking and to withstand the deep-sea pressures. Sometimes, prefabricated steel-mesh-concrete boxes of varying sizes were used instead of drums. As in the case of the steel drums, cement or concrete was mixed with the LLW to achieve the negative buoyancy necessary to assure sinking. This general design configuration was not intended to be permanent (U.S. General Accounting Office[5] – GAO, 1981; pp 2-9). It provided an estimated 10 years of radionuclide

---

[2]DOE self-regulates all LLW it generates and disposes of those wastes on-site at its facilities. DOE has developed a number of "orders" that address radioactive waste management. See DOE (1998). These orders do not have the legal enforcement mechanism of Federal regulations found in the *Code of the Federal Regulations* (CFR). Instead, DOE Orders are incorporated by reference into individual government contracts for vendors who operator the disposal facilities on behalf of DOE. DOE Order 435.1 (DOE,1999a) covers all HLW, LLW, transuranic (TRU) waste, and the radioactive components of mixed waste. Chapter IV of the implementing manual (DOE, 1999b) for DOE Order 435.1addresses the management of LLW. DOE does not classify LLW using NRC's Part 61 system. See NAS (1999 and 2003) and National Safety Council (2002).

[3]The AEA initially assigned the AEC the functions of both encouraging the use of nuclear power and regulating its safety. The AEC regulatory programs sought to ensure public health and safety from the hazards of nuclear power without imposing excessive requirements that would inhibit the growth of the industry.

[4]The NAS generally recommended that no 300-mile section of the coast line should contain more than three disposal areas and that adjacent disposal areas be separated by at least 75 miles. The NAS also had specific recommendations on the total quantity of activity disposed at any one location monthly as well as annually.

[5]In July 2004, the GAO was renamed the *General Accountability Office*.

containment[6] in the marine environment (NAS, 1959, p. 1).

More than 60 disposal sites were distributed between 5 major disposal locations in the Pacific Ocean, 1 in the Gulf of Mexico, and 11 in the Atlantic Ocean. The LLW was not evenly distributed among the sites; three sites received about 90 percent of the LLW, by volume. The number of LLW containers and the associated activity disposed is summarized in Table 1. Overall, it is estimated that about 95 percent of the containers disposed in the Atlantic and Pacific Oceans, and the Gulf of Mexico were 55-gallon drums (NAS, 1971; p. 36).

In 1970, the AEC ended its practice of disposing of LLW at sea. In the early 1980s, there was renewed interest in ocean dumping. In a report dated April 1984, the National Advisory Committee on Oceans and the Atmosphere (NACOA) recommended (pp. 6-7) that the present policy of excluding the use of the ocean for LLW disposal be reversed.

1.2   Land Disposal

In the 1960s, commercial interest in ocean disposal began to decline and had ended completely by 1970. One of the principal reasons was the adverse public reaction to polluting the ocean. The other motivation was economic. Ocean disposal was reported to cost as much as $48.75 per 55-gallon drum compared to $5.15 per drum for burial on land (Mazuzan and Walker, 1997; p. 367). For the reasons cited above, the AEC decided to endorse a new disposal policy permitting land burial using commercial disposal sites. Under this policy, it was envisioned that the private sector would identify sites with favorable geologic and meteorologic conditions and provide the same disposal service to commercial generators, but at lower cost. The intent was to geographically locate disposal sites in those regions generating the wastes. At the time, the Oak Ridge National Laboratory (ORNL) and the National Reactor Testing Station (now the Idaho National Engineering Laboratory) were the principal LLW handlers of wastes generated by then-AEC licensees. In contrast to ocean disposal, LLW generated there and at other at federal facilities was being disposed at about 16 major and lesser Federally-owned sites (NAS, 1976; p. 18-19).

Because of concerns about long-term institutional controls at potential commercially-operated sites, in January 1960, the AEC proposed that they be located on Government-owned land, regulated and licensed by the Government (Mazuzan and Walker, 1997; p. 366). As an interim measure, until a commercial disposal capability became available, the AEC decided to accept non-government LLW for disposal (Op cit., p. 367). In September 1962, the AEC authorized private firms to dispose of commercial LLW on land. The first privately-operated LLW land-burial service was near Beatty, Nevada, on State-owned land. The site was operated by Nuclear Engineering, Inc., one of the six firms already licensed by the AEC to commercially dispose of LLW at sea. At the time, licensing criteria specific to the disposal of LLW did not exist. The only applicable licensing criteria were the AEC's general regulations at 10 CFR 20.302(a) and (b).

Between 1962 and 1971, the six shallow-burial LLW disposal facilities were licensed and operated to dispose of the Nation's commercial LLW. Most of these facilities were located within the boundaries of or adjacent to a much larger federal reservation operated by the AEC

---

[6]With the exception of the radionuclides ^{90}Sr, ^{137}Cs, and possibly ^{60}Co.

Table 1. Summary of LLW Ocean Disposal Operations in the United States. Compiled using NAS (1959, p. 5), NAS (1971; p. 37), and NACOA (1984, Appendix C). However, the reader should note that questions have been raised in the past about the accuracy of past record-keeping and the accuracy of these statistics (Raubvogel, 1982, p. 23).

Water Body	Years of Disposal Activity	Number of Individual Disposal Sites	Number of Containers	Estimated Activity at Time of Packaging (Ci)
Atlantic Ocean	1951-1967	+24	34,203 [a]	79,482.9
Pacific Ocean	1946-1970	+34	56,261	14,980.5
Gulf of Mexico	< 1959	2	79	< 25

a. Includes unpackaged and liquid wastes.

(see Table 2). Four of the disposal sites – Beatty, Barnwell, Maxey Flats, and West Valley – were licensed by their respective host states through the Agreement State Program with the AEC (the predecessor to the NRC) under Section 274 of the AEA. The remaining two sites (Richland and Sheffield) were licensed by the AEC as their hosts had not become Agreement States at the time of licensing. Following the licensing of the second LLW disposal site (Maxey Flats), the AEC stopped accepting commercial LLW at its own facilities in May 1963.

The commercially-operated sites adopted the practice of near-surface, shallow-land burial (SLB) disposal technology adhered to at existing AEC facilities at the time. This disposal method relies on relatively simple engineering designs to isolate wastes from infiltrating water. The natural (geologic) characteristics of the site are the principal attenuators of any radioactive material that might be released to the accessible environment. There were no systematic site selection criteria or design requirements that could be used to establish the best mix of features necessary to contain and isolate the wastes. Disposal generally involved clearing and grading the land and excavating shallow unlined trenches (generally less than 50 feet deep) that would be used to receive the waste. At the time, LLW had no specific packaging requirements for disposal.[7] It was packaged in a variety of container types that were randomly dumped or stacked into the trenches. The waste was placed into the trenches generally on a first-come, first-serve basis. Trenches were then backfilled using materials removed during trench excavation, compacted, and graded to create an earthen mound cap necessary to prevent rainwater ponding and to promote runoff. The earthen cap was then re-seeded to grow a short-rooted protective grass cover. To preclude inadvertent intrusion, disposal sites were surrounded by a security fence. The working technical assumption behind this near-surface disposal method was that the nature and rates of natural processes acting on the earthen trench system would be sufficient to slow the movement of radionuclides from the disposal trenches to the accessible environment until they had decayed to acceptable background levels found in nature (EG&G Idaho, 1994; p. 4).

In 1973, the AEC asked the NAS to independently review the shallow-land disposal practices at its facilities. The AEC was particularly interested in identifying "... undesirable existing conditions and disposal practices ..." as well as identifying corrective actions, such as "... changes in current burial practices, changes in conditioning of [LLW] materials for burial, and special treatment of the ground prior to disposal ...." See Pittman (1973). The reason for this request was that routine monitoring at some of their sites had begun to reveal that the disposal trenches were not containing the wastes and radionuclides were being released (NRC, 1977a; p. 17). At the time, the AEC was also particularly concerned about the long-term management of the transuranic constituents of its wastes (NAS, 1976). In 1976, the NAS published its findings and recommendations following the review of solid LLW management practices at AEC facilities (later to become the Energy Research and Development Administration or ERDA[8]). Although the NAS found no serious deficiencies in past federal disposal practices, they did make numerous administrative as well as technical recommendations for it to consider.

---

[7]Department of Transportation (DOT) regulations for the transportation of radioactives wastes were first promulgated in 1979 (44 FR 1851).

[8]In 1974, the AEC was reorganized into the NRC and ERDA. See Section 2 of this paper for further details.

Table 2. Past and Current LLW Disposal Facilities. Taken from EG&G Idaho (1994), unless otherwise noted.

Site	Operational Period	Original Licensing Authority (year)	Status	Area (acres)	Disposal Volume (10e6 ft³)	Waste Form Characteristics			Comments
						By-product material (10e6 Ci)	Source Material (10e6 lbs)	Special Nuclear Material (lbs)	
Beatty, Nevada	1962 – 1992	AEC (1962)	Closed	80 (60)	4.7	0.64	4.0	605	A site adjacent to the now-closed LLW disposal facility is currently operated as a RCRA- and PCB-approved disposal facility.[b]
Maxey Flats, Kentucky	1963 – 1977	State (1962)	Closed	280	4.7	2.4	0.533	950	Designated as a EPA Superfund Site in 1986. Remediation completed in 1991.
West Valley, New York	1963 – 1975	State (1963)	Closed	3345 (22)[c]	2.5	1.3	1	125	LLW operations ceased in 1975 when burial caps leaked contaminated water.
Richland, Washington	1965 – present	AEC (1965)	Open	100[d]	13.6[e]	36.1[e]	13.5[f]	351[f]	Co-located within the Hanford nuclear reservation. Disposal site leased from the government. Disposal fees lower than Barnwell but higher than Envirocare.
Barnwell, South Carolina	1969 – present	State (1971)	Open	300	24.8[e]	12.8[e]	33.6[f]	6739[f]	Originally licensed for above-ground LLW storage. In 1971, LLW burial was approved. Highest disposal fees in the country.
Sheffield, Illinois	1968 – 1978	AEC (1967)	Closed	170 (20)	+3	3	0.06	126	Attempts to expand disposal capacity in 1975 were unsuccessful because of the detection of contaminated leachate, effectively ending site operations. In 1988, the Sheffield operator agreed to a 10-year monitoring plan with the State.
Clive, Utah	1991 – present	State (1991)	Open	540	25.0[g]	11.3[g]	NA	NA	Initially licensed to accept naturally-NORM. 1991 amendments permitted the disposal of LLW. Envirocare license has been amended 10 times to allow for the disposal of more types of LLW.

102

a. Actual disposal area, in parenthesis, smaller than area comprising disposal site.

b. In June 1988, the site operator, U.S. Ecology received a joint State-Federal RCRA permit to dispose of hazardous chemical wastes at a location adjacent to the LLW disposal site (Howekamp, 1996; p. 3). Pre-RCRA classified waste types had been disposed at this site since 1970. In 1978, the company also received EPA approval to operate a polychlorinated biphenyl (PCB) storage and disposal facility at the Beatty site.

c. Site owned by New York State includes multiple radioactive waste management areas.

d. Hanford nuclear reservation encompasses an area of 1000 acres.

e. Data for the period 1995-98 taken from Fuchs (1996, p. 6, 1997, p. 6; 1998, p. 6; and 1999, p. 7). Data for the period 1999 though June 2005 taken from MIMS.

f. Data only through 1994.

g. Data for period June 1992 through May 2005 taken from MIMS.

1.3   Early Performance Issues

After several years of operation, the West Valley, Maxey Flats, and Sheffield[9] sites began to encounter surface and/or ground water management problems. These problems coupled with other early LLW disposal practices resulted in the unexpected release and transport of radionuclides from the disposal sites. Key failure modes included waste container exhumation due to surface erosion, ground failures (subsidence) caused by inadequate waste container compaction, and the filling and subsequent release of contaminated leachate from disposal trenches resulting from infiltrating ground water (commonly known as the so-called "bath tub effect"). Because the disposal "units" were leaking, decisions were made to suspend operations and close the sites in the 1970s.

The remaining LLW sites had problems of a different type. The Beatty and Richland sites were temporarily closed in 1979 by the Governors of those States as a result of waste packaging violations and transportation safety issues. When the volume of waste shipped to the South Carolina site began to increase because of closures and interruptions at the other sites as well as a large increase in LLW generated following the Three Mile Island incident, there was concern that the facility would bear sole responsibility for the disposal of the Nation's commercial LLW. As a result, in 1979, the Governor of South Carolina ordered that waste acceptance operations be scaled-back by 50 percent over a 2-year period [see EG&G Idaho (1994)].

To address some of the past LLW site performance concerns as well as to develop geohydrologic guidelines that could be used to establish technical criteria for selecting, evaluating, licensing, and operating new LLW disposal sites, the U.S. Geological Survey (USGS) received direct LLW appropriations for the first time in 1975 (Schneider and others, 1982; p. 57) leading to the preparation of various reports. [See summaries in Trask and Stevens (1991), for example.] To reduce the potential for the environmental transport of radionuclides at disposal sites, the NAS (1976, pp. 67-68) independently recommended that arid sites in the west be considered based on the view that the geohydrologic settings there would be less complex and hence the view that performance could be more reliably predicted. Incidentally, the USGS reached this same opinion as early as 1974 recommending 17 types of earth-science information needed to predict the rate and direction of radionuclide transport. See Papadopulos and Winograd (1974). Other recommendations were made that some form of engineered barrier, working in concert with the natural system (geosphere), be integrated into future LLW facility designs (Battelle Memorial Institute, 1976; pp. 24, 48).

Technical issues notwithstanding, there still was the practical matter facing LLW generators that there had been a reduction in existing disposal access because of site closures and operational interruptions. Also, there was a mismatch in the geographic location of the remaining disposal facilities (mostly in the west) whereas most of the waste generation was in the east. The three remaining States with operating sites made it clear that they would not continue to accept all of the Nation's LLW (GAO, 1983; p. 7).

---

[9]Actually, contaminant transport was not discovered at the Sheffield site until after operators of the closed site attempted to re-open with expanded disposal capacity (EG&G Idaho, 1994; pp. 37-38).

2 CONGRESSIONAL ACTIONS

Congress abolished the AEC in the Energy Reorganization Act of 1974 (Public Law 93-438). The act placed the AEC's regulatory functions into the newly created NRC and placed the atomic energy promotional functions within ERDA, which were later absorbed into DOE following its creation in 1977.

2.1 The U.S. Nuclear Regulatory Commission and 10 CFR Part 61

The NRC began operations on January 19, 1975. The NRC (like the AEC before it) focused its attention on several broad issues that were essential to protecting public health and safety. Initially, NRC (and the AEC) regulated LLW using a collection of generic regulations specified in 10 CFR Parts 30 ("Rules of General Applicability to Domestic Licensing of Byproduct Material"), 40 ("Domestic Licensing of Source Material"), and 70 ("Domestic Licensing of Special Nuclear Material"). However, in response to the needs and requests expressed by the public, the States, Congress, industry and others, one of the earliest rulemaking efforts the Commission was to undertake was the development of a set of comprehensive requirements for licensing the land disposal of LLW.

In 1976, the GAO published a report that reviewed existing private and federal LLW disposal practices in light of the reported operational and performance irregularities identified at some disposal sites. Among other things, that review identified the need for studies and criteria to judge the suitability of LLW disposal sites as well as the need for standards to determine when releases from disposal sites reached unacceptable levels and corrective actions needed (GAO, 1976; pp. 19–21). In parallel to the GAO review, the NRC had formed a task force to examine its programs as well as those of the existing Agreement States that regulated commercial LLW disposal. Among the recommendations of that task force (NRC, 1977a; p. ii), was the need for NRC to "accelerate" the development of its LLW regulatory program. Shortly thereafter, the NRC (1977b) published a program plan that described the elements and schedules for implementing an integrated LLW program. This program plan included plans for the development of an environmental impact statement (EIS) and a yet-to-be-defined LLW regulation.

NRC began development of its LLW regulation in 1978 by relying on an extensive National Environmental Policy Act (NEPA) scoping process.[10] Early in that process, the Commission determined that comprehensive standards, technical criteria, and licensing procedures were needed to ensure public safety and long-term environmental protection in the licensing of new sites as well as the operation and closure of existing ones. The staff determined that the most viable approach to the regulation would be an "umbrella" regulation applicable to land disposal

---

[10]The National Environmental Policy Act of 1970 (Public Law 91-190) initially requires federal agencies to integrate environmental values into their decision-making processes by considering the environmental impacts of their proposed actions and reasonable alternatives to those actions. NEPA also requires that all Federal agencies prepare an EIS "for major actions significantly affecting the quality of the human environment." To meet this basic requirement, Federal and State governments, at all levels, now routinely prepare detailed EIS. In deciding to develop a LLW regulation, the NRC determined that the promulgation of Part 61 qualified as a major federal action.

The Council on Environmental Quality (CEQ) is responsible for developing regulations that implement NEPA. CEQ defines the scoping of an EIS at 40 CFR 1501.7 as "... an early and open process for determining the scope of issues to be addressed and for identifying the significant issues related to the proposed action...."

of most types of LLW. The challenge was that the regulation had to apply to a broad range of geologic/geomorphic conditions within the United States as well as apply to disparate waste streams. The other challenge was that early in the scoping process, the NRC staff determined that inadvertent re-entry into a LLW disposal area could not be precluded (NRC, 1980; 45 FR 13105). Consequently, the staff explored ways of classifying LLW for use in standardized exposure scenarios as a way of predicting potential doses to receptors. See Rogers (1979) and Rogers and others (1979). The staff also considered both generic and specific disposal methods in the context of an EIS that considered the costs, benefits, and impacts of a base-case disposal concept as well as alternative concepts. From those analyses and studies, performance objectives and technical criteria were proposed by the Commission in a draft regulation designated 10 CFR Part 61 (NRC, 1981; 46 FR 38081).

Following several years of development, the Commission issued a final Part 61 rule in December 1982. The regulation covered all phases of shallow, near-surface LLW disposal from site selection through facility design, licensing, operations, closure, post-closure, stabilization, to the period when active institutional controls end. The regulation also established the procedures, criteria, terms, and conditions on which the Commission would issue and renew licenses for the shallow-land burial of commercially-generated LLW (see Section 5 of this report). Among other things, Part 61 at §61.55, introduced a three-tier waste classification system for LLW based on the concentrations of the longer-lived radionuclides. These classes are designated Classes-A, -B, and -C in ascending order of potential radiological hazard, and the regulation had specific design standards applicable to each class.

2.2 The Low-level Radioactive Waste Policy Act of 1980

At the same time the NRC established a LLW regulatory framework, Congress passed the Low-Level Waste Policy Act of 1980 (LLWPA – Public Law 96-573). This act set forth a Federal policy that LLW disposal was best handled on a regional basis. The Act made States responsible for disposing of their own LLW generated within their borders,[11] and encouraged the States to form interstate compacts and establish regional disposal sites rather than establishing 50 separate disposal sites. The act was passed in response to policy recommendations from several states[12] and state-supported organizations, including the National Governors' Association and the National Conference of State Legislatures, intended to address past and present LLW management issues. The other key provision of LLWPA was that compacts were allowed to exclude LLW generated outside their borders.

Following passage of the act, the States began to enter into negotiations to form the required compacts. The States were generally committed to the compact arrangement. Shortly following its passage, 40 States had entered into agreements or were negotiating to form seven required compacts (GAO, 1983, p. 16). However, in its review of the compact-forming agreement process, GAO observed that the agreement process was "slow and drawn-out" despite having been devised by Congress (see Table 3). The GAO also observed that only

---

[11] By January 1, 1986, except for LLW generated by the Federal Government.

[12] Washington State, in conjunction with Nevada and South Carolina, sought passage of LLWPA because of the imbalance between the volumes of LLW those states were generating and the wastes they were receiving for disposal from outside their respective states. See Washington State Department of Health (2004, p. 43).

Table 3. Administrative Process for Establishing LLW Compacts. Taken from GAO (1983, p. 10). Compacts formed through this process are described later in Table 6 of this paper.

Step	Description of Activity
1	States negotiate among themselves to form regional Interstate Compacts of two or more states. [a]
2	Once formed, proposed Interstate Compacts draft Interstate Compact Agreements.
3	Drafted Interstate Compact Agreements are approved by the state legislatures and signed by the Governors in each state participating in the Interstate Compact.
4	Ratified (approved) Interstate Compact Agreements are to be approved by a majority of both Houses of Congress.
5	Following Congressional approval, each Interstate Compact is to form a Commission to administer the compact agreement. [b, c]

a. Alternatively, if a State chooses not to participate in the Interstate Compact process, it must indicate its intent not to do so. States deciding to act alone to meet their own LLW disposal needs still need to undertake the process steps outlined in Footnote b.
b. Once formed, the Interstate Compact Commission is responsible for ensuring that its member states (i) screen the region defined by the Interstate Compact to identify candidate disposal sites, (ii) select a preferred site and perform the required environmental assessment, (iii) prepare a LLW license application, and (iv) construct and operate the disposal facility, once the license application is approved.
c. Compacts can choose to defer the site selection/license application development process to a private entity.

three of the tentative compact regions had operating disposal sites, and those sites had been in existence before the passage at the act. GAO estimated that once a compact agreement had been entered into, it would take an additional 5 years before the disposal site was ready to receive LLW (*Op cit.*, pp. 20-21). Nevertheless, despite the progress being made GAO (*Op cit.*, p. 15) concluded that no new disposal sites would be operating until sometime after 1988, 2 years after the Congressionally mandated dated 1986.

When it became apparent that the deadline for operating new disposal sites would not be met, decision-makers recognized that adjustments to the existing LLW act were needed. Moreover, the three States with operating disposal sites made it clear that they would not continue to accept all of the nation's LLW. But before Congress could amend the 1980 act, an "understanding" was necessary between Nevada, South Carolina, and Washington – the States with operating disposal facilities – and the 47 unsited States. Following negotiations, these three states agreed to continue to receive out-of-State wastes for an additional 7 years, subject to certain conditions which where later reflected in the 1985 amendments to the act (NRC, 1989c p. 13).

2.3   The Low-level Radioactive Waste Policy Amendments Act of 1985

On January 15, 1986 Congress passed the Low-level Radioactive Waste Policy Amendments Act of 1985 or LLWPAA (Public Law 99-240). LLWPAA extended the original January 1, 1986, deadline to develop new disposal facilities by 7 years to January 1, 1993. Because new LLW disposal facilities were expected to be operational by the 1993 date, the existing States with operating LLW disposal facilities had the right, at that time, to decline receiving LLW from outside of their respective compacts. In exchange, the unsited States and regions were required to meet newly-established milestones and deadlines (see Table 4).[13] If States failed to comply with the specific LLWPAA milestones, the three States operating disposal facilities were authorized to deny disposal access those states in violation of the milestones. LLWPAA also included the following provisions:

- the establishment of financial penalties on waste disposed of at existing disposal facilities if certain milestones were not met.

- making the Federal Government responsible for disposing of commercial LLW exceeding Part 61 Class-C concentration limits.

- specifying which categories of LLW were exempt from LLW disposal facilities.

In passing the act, Federal agencies were given expanded responsibilities in the area of LLW (see Table 5). Specific new responsibilities were also assigned to DOE and the NRC. DOE was now required to do the following:

- dispose of GTCC-designated wastes,

---

[13]In a 1992 decision (New York vs. United States et al. – 505 U.S. 144) , the U.S. Supreme Court struck down the "take title" provision requiring that States must take title to their LLW if a disposal facility were not available by 1996.

Table 4. Milestones and Deadlines Defined by the LLWPAA.

Milestone Date	LLWPA Requirement
By July 1, 1986 ...	Each State shall join a regional compact by ratifying compact legislation or, by the enactment of legislation or the certification of the Governor, indicate its intent to develop its own LLW disposal facility.
By January 1, 1988 ...	Each compact region or the host State in which its LLW disposal facility is to be located shall develop a siting plan for such a facility providing detailed procedures and a schedule for establishing a facility location and preparing a facility license application and shall identify a developer to implement such plan.
	Each non-sited compact region shall identify the State in which its LLW disposal facility is to be located, or shall have selected the developer for such facility and the site to be developed, and shall identify a developer to implement such plan.
By January 1, 1990...	Each State (or the designated disposal facility developer) shall have submitted a complete application (as determined by the NRC or the appropriate agency of an agreement State) for a license to operate an LLW disposal facility or, in lieu of the license application, the Governor's written certification to the NRC, that such State will be capable of providing for, and will provide for, the storage, disposal, or management of any LLW generated within such State and requiring disposal after December 31, 1992, and include a description of the actions that will be taken to ensure that such capacity exists.
By January 1, 1992...	A complete application (as determined by the NRC or the appropriate agency of an Agreement State) shall be filed for a license to operate an LLW disposal facility within each non-sited compact region or within each non-member State.
By January 1, 1993...	Each State (or its compact region, where applicable) is expected to have provided a disposal facility for all the LLW it generates, and disposal rights at the three existing disposal facilities (Barnwell, Beatty, and Richland) will end.
	If a State (or, where applicable, a compact region) is unable to provide a disposal facility for its LLW, those States in the compact region shall, upon the request of the LLW generator or owner, be obligated to take title to and possession of the waste, or assume financial liability for costs associated with its storage and maintenance.
	If a State (or, where applicable, a compact region) is unable to provide a disposal facility for its LLW, the State (or States) will have to forfeit rights to rebates of previous surcharge payments made by LLW generators (or owners) because of the State's failure to meet earlier LLWPAA milestones.
By January 1, 1996... *	If a State (or, where applicable, a compact region) is unable to provide a disposal facility for its LLW, those States in the compact region shall, upon the request of the LLW generator or owner, be obligated to take title to and possession of the waste.

* In 1996, the US Supreme Court found that this provision of the 1985 Act was unconstitutional.

Table 5. **Federal Responsibilities for the Management and Disposal of Commercial LLW.** As defined by various Federal statutes.

Agency	Responsibility
Department of Energy	Overall lead agency for national planning of commercial LLW management and disposal. Assist in the forming of Interstate Compacts and establishing site selection procedures. Also undertake (or sponsor) research and development in the area of LLW disposal technology, and transfer that technology to the private sector.
Department of Transportation	Regulating waste containers, transportation vehicles, and other interstate aspects of LLW transport. [a]
Environmental Protection Agency	Establishing overall federal radiation protection guidance and environmental standards. [b]
Geological Survey	No basic responsibility for the management of LLW. Conduct basic research in the geological sciences and develop basic data for application in the development of disposal criteria. Also provide technical advice in the assessment of specific disposal sites.
Nuclear Regulatory Commission	Regulating and licensing the commercial and non-defense governmental use of source, by-product, and special nuclear material, including the licensing of commercial LLW disposal facilities.

a. Through a Memorandum of Understanding, the NRC and DOT have delineated their respective responsibilities for the transportation of radioactive wastes. The NRC regulates packaging for wastes containing high amounts of radioactive materials to assure safety and safeguards during transportation. DOT regulates all other aspects of radioactive waste transportation.
b. See Section 6.4.1 of this paper for more information.

- manage the collection of and disbursal of LLWPAA-levied surcharges[14],

- provide financial and technical assistance to the States and compacts, and

- generate certain status reports on the management of national LLW inventories.

For its part, the NRC was now required to the following:

- review LLW disposal facility license applications,

- develop standards and procedures for exempting certain LLW from disposal in licensed facilities,

- provide regulatory and technical assistance to Agreement States[15], and

- determine procedures for granting emergency access to LLW facilities for wastes generated in other regions[16].

Section 10 of LLWPAA also required that NRC establish standards for determining when radionuclides are present in waste streams in sufficiently low concentrations or quantities as to be "below regulatory concern" or BRC, and therefore not subject to NRC regulation. As early as February 1980, the staff indicated it intent to establish a *de minimis level* [17] for commonly-used, short-lived radioisotopes. In August 1986, the Commission published its proposed policy statement outlining its plans to establish certain new BRC rules and procedures (NRC, 1986a; 51 FR 30839). The Commission proposed that if radioactive materials did not expose individuals to more than 1 millirem per year (mrem/yr) or a population group to more than 1000 person-rem per year, they could be eligible for an exemption from full-scale regulatory control. However, this exemption would not be granted automatically; the NRC would consider requests from licensees that met the dose criteria through its rulemaking or licensing processes. The Commission intended that its BRC policy would apply to consumer products containing small amounts of nuclear materials and other sources of very low levels of radiation such that those types of wastes could safely be disposed in sanitary land-fills. The policy was also to provide a framework for making future exemption decisions and reviewing previous exemptions by which small quantities of low-level radioactive materials could be largely exempted from existing

---

[14]"Surcharges" were financial penalties imposed by DOE on waste generators if certain LLWPAA milestones were not met. These penalties were in addition to the basic disposal charges imposed by the disposal facility operator.

[15]Under Section 274 of the AEA, the NRC can relinquish portions of its regulatory authority to license and regulate byproduct materials (radioisotopes), source materials (uranium and thorium), and certain quantities of special nuclear materials to the States. The mechanism for the transfer of NRC's authority is an agreement signed by the Governor of the State and the Chairman of the Commission, in accordance with Section 274b of the act. "Agreement States" therefore are those States whose Governors have entered into such limited agreements with the Commission. At present, there are 33 NRC Agreement States.

[16]Promulgated as 10 CFR Part 62 (NRC, 1989a; 54 FR 5409).

[17]A *de minimus level* is one in which the radioactivity in the waste is sufficiently low that the waste can be disposed as ordinary, non-radioactive trash (NRC, 1980; 45 FR 13106).

regulatory controls (NRC, 1986b; p. 1). The Advisory Committee on Nuclear Waste (the ACNW or the Committee) provided two sets of comments to the Commission on the proposed BRC Policy in 1988. The NRC's proposed BRC policy was received unfavorably by both Congress and the public. As a result, it was officially withdrawn by the Commission in June 1993 (Walker, 2000; p. 120).

3   EFFORTS TO SITE NEW LLW DISPOSAL FACILITIES

The objectives of the LLWPA and LLWPAA were to provide for more LLW disposal capacity on a regional basis and distribute the responsibility for the management of LLW equitably among the States. In response to these two acts, by 1998, 44 states entered into 10 Interstate compact agreements. Compact membership varied from two to eight States per compact. As part of the compact agreement process, host States for the future disposal facilities were agreed-to and site-screening commenced. For those who already had not done so, designated host States entered into Agreement State programs with the NRC, and subsequently developed the regulatory and technical capabilities necessary to administer their respective programs. By definition, this would have included developing a regulatory framework compatible with the requirements of Part 61 and other NRC guidance (see Section 5 of this report).[17] In most cases, host states assigned the responsibility for implementing their respective programs to existing State agencies or created new or quasi-State authorities. Two regional compacts (Nebraska and California) delegated the disposal facility development responsibilities to private sector firms while retaining the regulatory functions.

As a result of these efforts, 7 out of 10 of the regional compacts were able to meet the first three milestones of the 1985 act leading to the submission of license applications. Regulatory authorities in four states (California, Illinois, Nebraska, and Texas) received license applications requesting authorization to construct new disposal facilities. California, however, the host State for the Southwestern Interstate Compact, was the only state able to proceed sufficiently in the licensing process to authorize the issuance of a construction authorization (see Table 6).

Despite these overall efforts, none of the States or compacts have been able to successfully develop new LLW disposal facilities. In their 1989 review, OTA found that some States enacted bans to legally restrict SLB disposal even though Federal regulations found that particular disposal method technically sound. Other issues cited by OTA included the rising costs of LLW disposal (at the time of the study, it had trebled in 20 years), and the management of mixed wastes. In California a contingent construction authorization for a new facility (Ward Valley) was granted by the State but the land transfer from the federal to the State government was never completed, effectively ending the facility's start up. LLW generators continue to rely on the existing disposal sites. Only one new disposal facility has actually been licensed – the Envirocare LLW disposal facility in Clive, Utah – was achieved outside of the LLWPAA framework.[18] Citing industry sources, the GAO (2004, p. 9) reports that national expenditures on various disposal facility development efforts since the passage of the LLWPAA may have reached approximately $1 billion. In its 1999 review of the national LLW program, GAO (p. 5)

---

[17] The Envirocare facility is located in Clive Utah and was initially licensed by the Utah Department of Environmental Quality to accept naturally-occurring radioactive waste or NORM (i.e., uranium mill tailings) for disposal. In 1991, the Envirocare license was amended by the State to permit the disposal of Class-A LLW, including mixed wastes, from all states except those in the Northwest Interstate Compact. On November 1, 1999, the operators of the Envirocare facility submitted a license amendment to the State to allow it to receive and dispose of containerized Classes-A, -B, and -C LLW.

[18] Current Commission regulations regarding NRC's relationship with the Agreement States are contained in 10 CFR Part 150 ("Exemptions and Continued Regulatory Authority in Agreement States and in Offshore Waters Under Section 274").

Table 6. LLW Compacts and LLWPA Milestone Status. Host state for the disposal facility is designated in bold type. "C" means completed disposal facility development milestones. Compiled using GAO (1992, 1999, and 2004).

Interstate Compact Region	Member States	Compact Formed	Select Site	Submit License Application	License Application Approved	Operate Facility	Comments
APPALACHIAN	Delaware Maryland **Pennsylvania** West Virginia	1985C	see Comments	—	—	—	Voluntary siting process suspended in 1991 because no municipality volunteered to host the disposal site.
CENTRAL	Arkansas Kansas Louisiana **Nebraska** Oklahoma	1982C	1989C	1990C	see Comments	—	A 1998 Nebraska denial of an application to construct was overturned in April 1999 by a US district court. In May 1999, Nebraska legislature voted to withdraw from Central Interstate compact. In 2004, a Federal appellate court ruling affirmed an earlier Federal district court decision that Nebraska, as a designated host State, is liable for $151 million in damages for reneging on its obligations to the Central Interstate compact to build a disposal facility by denying a license application for reasons not related to the merits of the initial application.
CENTRAL MIDWEST	**Illinois** Kentucky	1984C	1991C	1991C see Comments	—	—	In 1992, the Illinois legislature rejected conclusions of an earlier siting decision, effectively ending the license application review process. Since then, a new siting review process has been established as well as a cost-benefit analysis to determine whether a disposal facility should be built based on current LLW volumes.
MIDWEST	Indiana Iowa **Michigan*** Minnesota Missouri Ohio Wisconsin	1982C	see Comments	—	—	—	In 1997, the Midwest Interstate Compact Commission decided to suspend siting process noting that certain waste management action had taken place reducing the volumes of LLW being generated within the compact.

Interstate Compact Region	Member States	Compact Formed	Select Site	Submit License Application	License Application Approved	Operate Facility	Comments
NORTHEAST (later renamed ATLANTIC)	Connecticut[b]	1985/2001C	see Comments	—	—	—	State legislature terminated siting efforts in 1992 citing the availability of out-of-state disposal capacity.
	New Jersey[b]		see Comments	—	—	—	State siting board terminated siting efforts in 1992 also citing the availability of out-of-state disposal capacity.
	South Carolina		C	C	C	1969 C	In 2001, South Carolina legislation restricted the use of the Barnwell disposal facility to generators in the three-member Atlantic Interstate compact after mid-2008.
NORTHWEST	Alaska Hawaii Idaho Oregon Montana Utah **Washington** Wyoming	1985C	C	C	C	1965 C	The compact's regional disposal facility is the existing Richland (Washington) facility.
ROCKY MOUNTAIN	Colorado **Nevada** New Mexico	1985C	C	C	C	1965 C[e]	Since the closure of the Beatty site, the compact has contracted with the Northwest Interstate compact to dispose of LLW at the existing Richland facility.
SOUTHEAST [d]	Alabama Florida Georgia Mississippi **North Carolina** Tennessee Virginia	1985C	see Comments	—	—	—	South Carolina withdrew from compact in 1995. State siting board terminated operations in 1997 because insufficient funding. In 1999, North Carolina withdrew from the compact (GAO, 1999; p. 72). In 2000, North Carolina joined the re-named Atlantic Compact (GAO, 2004; p. 28).

Interstate Compact Region	Member States	Compact Formed	Select Site	Submit License Application	License Application Approved	Operate Facility	Comments
SOUTHWESTERN	Arizona California North Dakota South Dakota	1985C	1988C	1989C	1993C	see *Comments*	From 1993-96, the Secretary of the Interior deferred making a land-transfer decision necessary to construct and operate the State-approved Ward Valley site while a number of technical review and administrative activities were underway by the government and the NAS. See GAO (1977). In a 1999 court decision brought on by California, it was found that the Federal government was not required by Federal law to transfer (sell) the land. Since that decision, there have been no additional siting activities by the State.
TEXAS	Maine Texas Vermont	1998C	1992C	1992C see *Comments*	—	—	In 2003, the Texas legislature designated a geographic area in the State as acceptable for a new disposal facility, and the host state's regulator developed a license application process for this facility.
UNAFFILIATED	District of Columbia New Hampshire Rhode Island Puerto Rico	These States do not intend to build LLW disposal facilities. They will seek storage and disposal arrangements with other States.					
	Massachusetts	n/a	see *Comments*	—	—	—	In 1995, the state hired a contractor to conduct a state-wide screening process. In 1996, the process was terminated because of renewed access to the Barnwell disposal facility (GAO, 1999; p. 76).

Interstate Compact Region	Member States	Compact Formed	Select Site	Submit License Application	License Application Approved	Operate Facility	Comments
	New York	n/a	see Comments	---	---	---	In 1988, the state's independent siting commission conducted a multi-step screening process to identify candidate sites for evaluation as LLW disposal sites. In its independent review of the site selection process, GAO (1992) found that the state did not adhere to its administrative procedures for selecting candidate sites. The Governor later suspended the siting process. In 1995, the state legislature declined to fund the siting commission (GAO, 1999; p. 76).

a. Michigan expelled from compact in 1991 for not acting in good faith to locate an acceptable disposal site. Ohio is the alternate host State.
b. Originally intended as dual host states in 1985 as part of the Northeast Interstate compact region. In 2001, the two States, along with South Carolina, formed the Atlantic Interstate compact region.
c. The Beatty facility provided disposal service to the Rocky Mountain Interstate Compact until 1992.
d. The Barnwell site in South Carolina provided the Southeast Interstate Compact region with disposal service until 1995, at which time it withdrew from the compact.

identified some common reasons for the lack of success in providing new disposal facilities. The reasons included the following:

- the controversial nature of nuclear waste disposal and public opposition to the siting of new LLW disposal facilities.

- the declining volumes of LLW being generated as a result of waste minimization and processing into safer forms.

- the high costs associated with the siting, licensing, constructing, and operating of new disposal facilities.

- the continued availability of existing disposal capacity.

- the consideration of alternatives to disposal – e.g., assured isolation.

4   CURRENT PROGRAM STATUS

In the mid-1990's, the NRC significantly scaled-back its LLW program for budgetary reasons. At the time, the actions were justified as the NRC already had a regulatory framework in-place sufficient to review a Part 61 license application,[19] and the Commission had relinquished its licensing authorities to those host states with a lead role in developing new LLW disposal facilities. Another factor cited was the lack of national progress in siting new disposal facilities.

To keep abreast of national LLW developments under the current reduced program, the staff has done several things. For example, the staff regularly monitors developments within the national program by attending regular meetings of the *Low-Level Radioactive Waste Forum*.[20] The staff has also performed several specific tasks, as directed by the Commission. They include efforts to improve the transparency of NRC decision-making as it relates to Section 20.2002 requests[21] and determine whether depleted uranium needs to be added to the Part 61 waste classification system.[22]

Consistent with earlier Congressional direction, DOE established a *National Low-Level Radioactive Waste Management Program* to develop and make available useful information concerning LLW management. Under contract to DOE, the operating contractor for the Idaho National Engineering and Environmental Laboratory prepared technical reports covering many LLW areas – e.g., SLB corrective measures (EG&G, 1984), LLW laws and administration (EG&G, 1985), and environmental monitoring (EG&G, 1989).[23] From 1979 to 2000, annual State-by-State assessment reports were also prepared that provided information on the types and quantities of LLW (e.g., Fuchs, 1999). In 1986, DOE developed a computerized *Manifest Information Management System* (or MIMS[24]) to monitor the management of commercial LLW. This system later subsumed the annual State-by-State assessment reports series. In 2000,

---

[19]This framework is discussed in more detail in Sections 6.4.4 and 7.1 of this paper.

[20]Until 1985, representatives of the Governors worked to achieve the goals of the LLWPA through a committee of the National Governors' Association. After passage of the 1985 amendments, representatives of compacts and States established a separate organization, known as the *Low-Level Radioactive Waste Forum* (LLW Forum), to promote the objectives of the new Federal law and the compacts. In 2001, the LLW Forum became an independent nonprofit organization.

[21]From time to time, the Commission receives requests to permit the disposal of small quantities of low-activity radioactive materials, on site, at existing NRC-licensed facilities. Disposal exemptions to Part 61 are allowed under NRC's regulation at §20.2002 ("Method for obtaining approval of proposed disposal procedures") to Part 20 ("Standards for Protection Against Radiation"). Staff guidance regarding the on-site disposal of small qualitites of radioactive waste can be found in Goode and others (1986), Neuder (1986), Neuder and Kennedy (1987). The Commission can grant other types of disposal exemptions under §§ 20.2003 (sanitary sewer releases), 20.2004 (incinerator releases), and 20.2005 (biomedical waste releases).

[22]In a decision dated October 19, 2005, the Commission directed the staff to determine whether depleted uranium produced by uranium enrichment facilities warrants consideration under §61.55(a) of NRC's waste classification tables. See Diaz and others (2005).

[23]Time limitations in the development of this paper did not permit a review on the DOE-sponsored literature.

[24]The MIMS webs site can be found at *http://mims.apps.em.doe.gov/*.

Congress stopped appropriating money for the DOE's national LLW program with the exception of the funds necessary to maintain MIMS. In its 2004 evaluation of the National program, GAO (pp. 14–16) found shortfalls in the quality of the MIMS data and recommended that the NRC take responsibility for generating the required reports. The GAO was particularly concerned that the unreliability of the data would make it difficult to forecast future disposal needs for all classes of LLW.

In its 1994 and 2000 reports described earlier, the GAO assessed three management options to address concerns about limited or no disposal access for LLW generators. The three options suggested were:

- retain the existing compact approach and allow it to adopt to the changing LLW situation;

- repeal the existing LLW legislation and allow market forces to respond to the changing LLW situation; or

- use existing DOE facilities for the disposal of commercial LLW.

Most recently, in November 2005, the GAO was directed by congress to report on approaches to improve the management of LLW within the United States. This examination is expected to include a review of best practices internationally (GAO, 2005).

4.1   Recent Disposal Facility Developments[25]

The Beatty LLW disposal site was permanently closed in 1992 by order of the State's Governor. The site is currently operated as a RCRA and PCB waste disposal facility. The nation's only remaining disposal facilities are at Barnwell, Richland, and Envirocare. Only the Envirocare facility receives mixed LLW. The Barnwell facility presently receives Classes-A, -B, and -C LLW. In 2000, the South Carolina Legislature restricted disposal access to the facility to only members of the Atlantic Interstate Compact after mid-2008. In 2001, a license amendment was approved by the state regulatory authority in Utah to allow the Envirocare facility to dispose of Classes-B and -C LLW. Under state law, approval of the legislature and governor are now required before Classes-B and -C waste can be received (GAO, 1994; p. 33). In late 2005, the governor voiced his opposition and placed a moratorium on the acceptance of these wastes.

Attempts are also underway to site a LLW disposal facility in Andrews County, Texas, by Waste Control Specialists (WCS), LLC. The WCS facility is located on 14,400 acres. More than 1340 acres is currently permitted to treat and dispose of RCRA waste and Toxic Substances Control

---

[25]Taken primarily from GAO (2004).

Act (TSCA) materials. The Andrews WSC site is also permitted for GTCC LLW storage, PCB-contaminated waste treatment, storage and land disposal, AEC Section 11e.(2) waste storage, and NRC exempt and exempt-mixed waste land disposal, including selected NORM waste. In 2003, the Texas Legislature passed legislation that allows a private entity to make an application for an NRC Part 61 LLW disposal site (Lauer, 2003; p. 13). In August 4, 2004, WSC submitted a license application to the Texas Commission on Environmental Quality to construct a near-surface LLW disposal facility. That license application is currently under review. The State regulatory agency has found the license application to be complete and expects to publish the draft license and hearing notice in mid-2006.

4.2   Assured Isolation

As an alternative to permanent (geologic) disposal, the concept of *assured isolation* has been proposed by Newberry and others (1995). Unlike the prevailing Part 61 disposal concept, assured isolation was considered to be a more publicly-acceptable alternative for it calls for caretaker oversight to allow for the indefinite storage of LLW in an engineered facility until such time that the waste no longer poses a radiological hazard. Conceptually, assured isolation is envisioned to be simpler from a technical standpoint in that structures, systems, and components (SSCs) of the *assured isolation facility* (or AIF) do not have to be designed to maintain their intended safety functions after the facility is closed, as would be the case with permanent disposal. AIF is envisioned to effectively function as a monitored storage system, with the capability for regular inspections and maintenance of SSCs to ensure that the wastes are being contained and, if necessary, retrieved. Unlike a Part 61 disposal facility, which has several performance objectives, applying to both the pre- and post-closure operational phases, the AIF would only have to meet the Part 20 worker safety standards so long as the wastes remained hazardous,. Because the site (geosphere) is no longer a consideration in the performance of the system, the need for detailed site characterization, complex performance assessment analyses, and the development of a long-lived waste package is obviated. Also see Newberry and others (1996).

In a September 2002 SRM, the Commission directed the NRC staff to explore interest in the assured isolation concept and develop a rulemaking plan that could be used to provide a foundation for a Commission decision on whether to develop such a rule. The need for a rulemaking plan was prompted by the development of a draft AIF regulation by the State of Ohio, and the State's subsequent request for NRC to review and comment on that draft regulation. At the time, at least 5 other states were contemplating similar regulations. The Ohio rule is now the only AIF regulation currently in effect. For its part, the NRC has no regulations or criteria for the design and operation of an AIF. To ensure consistency with any future state regulations, the staff has previously recommended the development of an AIF rule. However, before proceeding to develop such a rule, the staff surveyed the States, Interstate Compacts, and industry representatives to determine how widespread the support was for an NRC regulation in this area; responses to that survey suggested only limited interest. See SECY-03-0223 (NRC, 2003). Should NRC promulgate an AIF regulation, Ohio and any other states with similar regulations would be required to modify those regulations to be consistent with NRC's, based on the Commission's AEC authorities. In a January 2004 SRM, the

Commission has directed the staff to defer action on the development of an AIF rule and annually review the need for further action in this area.

The literature does not indicate if there has been a detailed comparison between the AIF and the prevailing Part 61-based disposal concepts. In a 2005 review of DOE's LLW management programs, the GAO (2005) recommended the use of life-cycle cost analyses to evaluate competing LLW management alternatives.

4.3   Stakeholder Views

In addition to the National program reviews by the GAO and OTA, some LLW stakeholder organizations and entities have prepared position papers expressing their views on various matters related to the management of commercial LLW. Some of these position papers also call for regulatory changes to NRC's LLW regulatory framework. A Internet search summarized in Table 7 indicates that there are several published position papers. These position papers provide different perspectives and sometimes conflicting positions on stakeholder views. No attempt has been made to summarize the opinions expressed. The reader is referred to the individual papers to better understand the respective views of the organizations who have prepared these papers.

Table 7. Availability of Stakeholder Position Papers on LLW.

Organization/Entity	Internet Homepage	LLW Policy Statement	
		Title	Date
INTERSTATE LLW COMPACTS			
Southeast	http://www.secompact.org/	"Management of Low-Level Radioactive Waste"	November 30, 2005
STAKEHOLDER ORGANIZATIONS			
American Nuclear Society	http://www.ans.org/	Disposal of Low-Level Radioactive Waste – Position Statement No. 11	November 2004
California Radioactive Materials Management Forum (Cal Rad Forum)	http://www.calradforum.org/	"A National Solution for a National Problem"	2003
Council on Radionuclides and Radiopharmaceuticals (CORAR)	http://www.corar.org/	"Council on Radionuclides and Radiopharmaceuticals Position Paper on Low-level Radioactive Waste Disposal"	April 6, 2001
Health Physics Society	http://www.hps.org/	"Low-level Radioactive Waste Management Needs a Complete and Coordinated Overhaul"	September 2005 (revision)
League of Women Voters	http://www.lwv.org//AM/Template.cfm?Section=Home	"Environmental Protection and Pollution Control" [general subject of LLW management]	July 5, 2005
LLW Forum	http://www.llwforum.org/	"Management of Commercial Low-Level Radioactive Waste"	September 22, 2005
National Governor's Association	http://www.nga.org/portal/site/nga	NR-19 Policy Position: "Low-Level Radioactive Waste Disposal Policy"	February 26, 2004
National Mining Association	http://www.nma.org/	"The National Mining Association's and the Fuel Cycle Facilities Forum's White Paper on Direct Disposal of Non-11e.(2) Byproduct Materials in Uranium Mill Tailings Impoundments" [includes a discussion of LLW]	No date

PART II: NRC's LLW REGULATORY FRAMEWORK

5   INTRODUCTION

Without exception, all past case studies of LLW disposal pointed to the need to improve its management to ensure that the wastes, once disposed, would not create a public health hazard. Applied to the disposal of LLW, this meant not only protecting workers and the public during the operational phase of waste disposal but also assuring that once a facility was closed, the disposal "system" contained the waste for a period of time sufficient to ensure that it no longer posed a hazard.

In response to the needs and requests of the public, the States, industry and others, the Commission promulgated specific requirements for licensing the near-surface land disposal of commercial LLW[26] at Part 61. These requirements were developed during the 5-year period from 1978 to1982 following the 1977 recommendations of an internal NRC task force (NRC, 1977b). NRC's final commercial LLW disposal regulation was published in the *Federal Register* on December 27, 1982 (47 FR 57446). The rule applies to any near-surface LLW land disposal technology. This includes SLB, engineered land-disposal methods such as below-ground vaults (BGVs), earth-mounded concrete bunkers (EMCBs), and augered holes. The regulation emphasizes an integrated-systems approach to LLW disposal, including consideration of site selection, site design and operation, waste form, and disposal facility closure. To lessen the burden on society over the long periods of time contemplated for the control of radioactive material, Part 61 emphasizes passive rather than active systems to minimize and retard releases to the environment. Various subparts of the rule cover general provisions and procedural licensing aspects, as well as those subparts covering the performance objectives, financial assurances, State and Tribal participation, and records, reports, tests and inspections. Existing LLW disposal sites were not required to conform to the Part 61 requirements, although many of the features of the regulation were incorporated as license conditions for existing facilities.

Since 1983, the NRC staff has developed several documents intended to aid in the implementation of Part 61. Foremost among these are NUREG-1300 – "Environmental Standard Review Plan for the Review of a License Application for a Low-Level Radioactive Waste Disposal Facility (Environmental Report)" (NRC, 1987); NUREG-1199 – "Standard Format and Content of a License Application for a Low-Level Radioactive Waste Disposal Facility" (NRC, 1991a); and NUREG-1200 – "Standard Review Plan for the Review of a License Application for a Low-Level Radioactive Waste Disposal Facility" (NRC, 1994). NUREG-1199 details the necessary components and information needed in a license application for an LLW disposal facility required under Part 61. NUREG-1200 provides guidance on the process that the staff would use to review a Part 61 license application. Consistent with the requirement in the LLWPA to review a Part 61 license application within 15 months of its receipt, NUREG-1274 (Pittiglio, 1987) was prepared that describes the staff's approach to reviewing any potential license application. In issuing an Part 61 LLW disposal facility license, the NRC would be required to prepare and issue an EIS. NRC Regulatory Guide 4.18 (NRC, 1983b) and NUREG-

---

[26]LLW waste is defined in Part 61 the same way as it is defined in the LWPAA and the Nuclear Waste Policy Act of 1982, as amended [i.e., radioactive waste that is not classified as HLW, TRU waste, SNF, or byproduct material as defined in Section 11e.(2) of the AEA (i.e., uranium or thorium tailings and waste).]

1300 (NRC, 1987) provide guidance to the staff on what should be included in the EIS. Because of the key role quality assurance (QA) has played in the nuclear program, the NRC staff has also developed specific QA guidance for the LLW regulatory arena. NUREG-1293 (Pittiglio and Hedges, 1991) provides specific guidance on how to meet the Part 61 requirements.[21] NUREG-1383 (Pittiglio and others, 1990) provides QA guidance related to site characterization activities. Additional QA guidance for potential Part 61 applicants is provided in Chapter 9 of both NUREG-1199 and NUREG-1200.[22]

---

[21]The criteria described in NUREG-1293 are similar to the criteria contained in Appendix B of 10 CFR Part 50. Although Appendix B to Part 50 is not applicable to NRC's LLW disposal regulation, the criteria it contains are basic to any nuclear regulatory QA program.

[22]Section 8 of the LLWPAA also directed the NRC to identify and publish technical information for disposal methods other than SLB. NRC complied with this provision by publishing NUREG-1241 (Higginbotaham and others, 1986) and NUREG/CR-3774 [Bennett and others (1984), Bennett (1985), Bennett and Warriner (1985), Miller and Bennett (1985), and Warriner and Bennett (1985)]. In addition, the NRC revised NUREG-1199 and NUREG-1200 to address BGVs and EMCBs, in addition to SLB.

6 APPROACH TO DEVELOPING PART 61[23]

Before the promulgation of Part 61, there were no standards nationally or internationally defining what level of safety was necessary to protect the public from disposed LLW. The only comparable regulations in place that defined "safety" were the AEC generic criteria relating to occupationally-exposed workers during the operation of licensed nuclear facilities found at Part 20. These criteria define the maximum permissible levels of radiation in unrestricted areas. Although Part 20 does not contain technical criteria or standards specific to the disposal of licensed materials such as LLW, it was nevertheless used to license early LLW disposal facilities because the regulation was generally intended to protect both workers and members of the public.

Consistent with the staff's 1977 plan, the Commission published an *Advance Notice of Proposed Rulemaking* (or ANPR), in October 1978 (43 FR 49811), inviting advice, recommendations, and comments on the scope of the EIS the staff was developing in support of the new Part 61 regulation. The EIS proposed was not intended to be a generic EIS on LLW disposal vis-a-vis the NEPA process. Rather, it was intended to serve as the document that would provide the bases and record for Commission decisions on the requirements to be set out in the forthcoming regulation. To ensure that no viable LLW disposal alternatives would be overlooked, as part of the scoping process, the NRC sponsored a technical study (Macbeth and others, 1978) that was included as part of the 1978 ANPR. Also see Denham (1988).

The comments received by the Commission during the ANPR were used to scope and form the content of the draft EIS, designated as NUREG-0782 (NRC, 1981) as well as the preliminary draft regulation which was made available for public comment on February 28, 1980 (45 FR 13104). The draft regulation identified the licensing procedures, performance objectives, and technical requirements necessary for the licensing of LLW disposal facilities. The proposed regulation also reflected NRC's long-standing ALARA or "as low as reasonably achievable" regulatory principles. During the summer and fall of 1980, the Commission also sponsored four regional workshops to provide stakeholders with an opportunity to discuss the issues addressed in the proposed Part 61 rulemaking. The Commission received 36 comments from the public on the ANPR. The respondents strongly supported the Commission's development of specific standards and criteria for the disposal of LLW (NRC, 1981; 46 FR 38082). Among the comments received were specific recommendations that a system was needed for classifying or segregating the waste based on (radiological) hazard (46 FR 38082). After consideration of the information received, the Commission published its proposed Part 61 LLW regulation on July 24, 1981 (46 FR 38081). The NRC staff and one of its technical assistance contractors, ORNL, conducted a series of three symposia between 1981 and 1983 intended to examine technical issues related to the siting, design, and/or performance of LLW disposal facilities as well as the proposed draft Part 61 regulation. See Yalcintas and Jacobs (1982a) and Yalcintas (1982b and 1983).

---

[23]The purpose of this section is to provide some general background on the approach used to develop Part 61 and in doing so, highlight a few key issues considered important at the time. This summary is not intended to be exhaustive. NUREG-0782 and NUREG-0945 provide a more detailed account of this development process as well as the staff's and Commission's disposition of key issues related to that development.

The Commission received comments from 107 individuals, organizations, and entities on the proposed regulation. The general response to the proposed rule was considered favorable by the Commission (NRC, 1982b; 47 FR 57447). For the most part, the commenters were evenly split, either declaring explicit support of the rule and the Commission's proposed overall regulatory approach or, offering constructive comments on specific aspects of the proposed rule without taking a general position on the rule itself or offering to support, with reservations. No State group or existing LLW disposal site operator expressed opposition to the proposed rule. Only 15 commenters expressed outright opposition to the rule or some significant portion of it. As a result of the generally favorable comments received, the Commission finalized Part 61 in 1982 (47 FR 57446). To support publication of the final rule the staff also issued a final EIS – designated NUREG-0945 (NRC, 1982a), which contained a detailed analysis of the comments received on the draft EIS as well the decision bases and staff positions in support of the final regulation.

6.1   Elements of the LLW Regulation

The Part 61 regulation applies to any land disposal technology for commercial LLW. The regulation covers all phases of LLW disposal from site selection through facility design, licensing, operations, closure, and post-closure stabilization, to the period when active institutional controls end. The regulation also establishes the procedures, criteria, terms, and conditions on which the Commission would issue and renew existing licenses. The requirements emphasizes an integrated-systems approach to shallow land disposal, including consideration of site selection, site design and operation, waste form, and disposal facility closure. Because of the long periods of time contemplated for the control of radioactive material, Part 61 also emphasizes passive rather than active systems to minimize and retard releases to the environment. To provide flexibility in siting and designing disposal facilities, a LLW waste classification system was devised based on half-lives and concentrations of radioactive materials that are expected to be in the wastes. All commercial LLW classes are subject to minimum waste form characteristics.

The Part 61 regulation is organized into several subparts. Various subparts cover general provisions and procedural licensing aspects, as well as those subparts covering the performance objectives, financial assurances, State and Tribal participation, and records, reports, tests and inspections. In addition, the regulation specifies requirements that must be met by the waste generator, including requirements for waste form and content, waste classification, and waste manifests. See Appendix A for more details on major subject areas in the regulation.

As noted previously, the focus of the Part 61 regulation is on the long-term disposal of LLW, which was unique concept at the time. The Commission employed a top-down, integrated systems approach to developing it. It proposed performance goals (objectives) that accounted for both the short-term as well as long-term radiological exposures. The regulation was oriented towards overall performance goals defined in Subpart C that define the objectives (regulatory policies) to be achieved in waste disposal. The performance goals are supported by a narrow (minimum) set of prescriptive technical standards in Subpart D based on past operating experience judged to be important to meeting the overall performance goals. The intent of this regulatory approach was to allow some flexibility to LLW disposal facility developers, consistent with a particular geologic and/or geographic setting, in choosing advantageous siting and design features and operating practices necessary to achieve the performance goals (NRC,1981; 46 FR 38083). The Commission chose not to include too much specificity in setting the technical standards as that would require considerable amount of detailed knowledge about the spectrum of designs, techniques, and procedures for disposing of LLW. Alternatively, the Commission chose to provide prospective applicants flexibility in deciding how the performance objectives would be met.

Through the earlier scoping process, the site (geosphere) is considered to be part of the containment system which, in concert with specific design features (e.g., clay liners, engineered barriers), would slow the expected release of LLW to acceptably small quantities of radioactive material over time. The technical requirements apply to site suitability, specific features of the facility design, operations and closure, waste classification, waste form and certain institutional assurance measures. Requirements for environmental monitoring are also established in the post-closure phase to assess the overall system's performance. These minimum technical requirements were collectively deemed important to achieving successful waste disposal by the Commission based on past reviews and experience. By relying on multiple barriers, reliance is not placed on any one component of the LLW disposal system to ensure that the performance objectives are met. Rather, all components of the system, acting in concert, are intended to contain and isolate the wastes. This concept of multiple barriers is consistent with the Commission's traditional views regarding *defense-in-depth*[24] and aids in the decision-making for issuing a Part 61 license using the standard of *reasonable assurance*.[25]

6.2　Who Should be Protected and What Should the Level of Protection Be?

As noted earlier, the Commission's intent in promulgating Part 61 was to develop an umbrella regulation that addressed all phases of the LLW disposal cycle. This meant that the regulation had to be sufficient to cover disposal operations and closure as well as the long-term period of waste isolation.

The performance objectives defined in Subpart C were developed expressly for commercial LLW.[26] They define the overall level of safety to be achieved by disposal. Although the Commission's requirements at Part 20 were considered appropriate to existing types of nuclear facility operations, they were not considered appropriate for the long-term disposal of LLW (NRC, 1989; p. 7). The Part 61 performance objectives are intended to provide protection from normal disposal facility operations as well as longer-term protection from the release of radioactive materials after facility closure, including accidental exposures caused by inadvertent

---

[24]"Defense-in-depth" is more of an NRC design principle and operational philosophy rather than a regulatory requirement per se. One of the essential properties of the principle is the concept of employing successive physical barriers that provide redundancy to what is in this case a disposal system containing LLW. This principle applied to NRC's regulatory programs is discussed in more detail in NRC (1995, 60 FR 42622), NRC (1998b), and Powers (1999).

[25]Section 61.23 defines thee standards the Commission will use to determine if it can issue a Part 61 license application to operate a LLW disposal facility. In issuing any license, the Commission would apply the standard of *reasonable assurance*. Historically, the concept has been used by the Commission to describe the acceptability of information submitted in a license application that wold demonstrate that the licensed facility would perform as intended and in doing so, protect public health and safety. See Schweitzer and Sastre (1987, pp. 4–5).

[26]In the absence of applicable environmental radiation standards promulgated by EPA, the NRC developed the four performance objectives through rulemaking. See Section 6.4.2 of this paper for further information.

intrusion and waste exhumation, in which the intruder is unaware of the presence of the waste. The technical requirements in Subpart D are considered minimum requirements intended to help ensure compliance with the performance objectives.

The Commission was also concerned about the potential for inadvertent intrusion once institutional control of the site had ended and knowledge of the hazard ceased. In relying on near-surface disposal, there is the possibility for exposures to ionizing radiation resulting from man's efforts to reclaim a disposal site for productive use such as farming, housing, or natural resource development. Archeological activities and scavenging could also lead to waste exhumation. The staff recognized early in the EIS scoping process that because there was no basis for predicting these types of behavior, there was no way to guarantee that inadvertant human intrusion at the site would not occur in the future (NRC, 1981; 46 FR 38083). Consequently, the staff determined that future generations, in effect, should be afforded the same level of protection as the general population today. Although widely used today in the evaluation radioactive waste disposal systems,[27] the human intruder scenario was a unique concept at the time it was first proposed by the Commission.

In another type of inter-generational equity concern, the Commission also took the position that future generations should not bear the responsibility for managing wastes produced by past generations. To address this issue, the Commission took the position that the disposal facility, its components, and even certain types of LLW should be robust and recognizable for some minimum period of time into the future while the radiological hazard still exists so as to preclude the potential for releases into the environment (NRC, 1982; 47 FR 57457, 57459).

As a result of these considerations, during the rulemaking scoping process, the following performance goals found in Subpart C were proposed:

• Protect members of the public and occupationally-exposed workers during facility operation (at §§61.41 and 61.43).

• Protect inadvertent human intruders entering the facility once disposal operations ceased and the facility decommissioned (at §61.42).

• Assure the long-term physical stability of the disposal facility to obviate the need for long-term maintenance after decommissioning of the facility (at §61.44).

These performance goals effectively defined Commission's policy on who would be protected (and when) as the result of the operation of a reference LLW disposal facility. The first performance goal applied to short-term exposures associated with the pre-closure phase of facility operations. As noted earlier, the intent of this requirement was to ensure that LLW disposal facilities would be operated in conformance with the same standards for radiation protection that the Commission already applied to existing materials licensees. As a

---

[27]The deterministic modeling of human intrusion is now a widely-employed analytical technique that is used to evaluate the robustness of radioactive waste disposal concepts to the disruptive consequences of borehole drilling. See, for example, Charles and McEwen (1991), Nuclear Energy Agency (1991), Berglund (1992), and Wescott (2001).

consequence, this performance goal required compliance with existing Part 20 criteria for radiation exposure to workers.

With the update of Part 20 (NRC, 1991c) there are now two different bases for doses in §§61.41 and 61.43. The whole body and organ dose limits specified in §61.41 are based on the older system of dose calculation based on the methods documented in the *International Commission on Radiological Protection* (ICRP) *Publication 2* (ICRP, 1960). This system is based on the principles of maximum organ burdens and intakes so annual doses are limited to the maximums allowed for critical organs. Now Part 20 is based on ICRP *Publications 26 and 30*. See ICRP (1977 and 1979-88,[28] respectively). The principles in these reports are based on estimating doses for 50 years for intakes that occur in a year of practice and limiting exposures so that the assigned dose for intakes in that year do not exceed limits.

The practical result is that under the new system long lived radionuclides are more restricted that under the old system. In short a dose expressed in mrem/yr whole body using the concepts in §61.41 is not necessarily equivalent to 25 mrem total effective dose equivalent (TEDE) assigned to a year of practice using the concepts in Part 20 and therefore by greater if more long lived radionuclides are involved in internal exposures. The difference is greater if more long lived radionuclides are involved in internal exposures.

Although appropriate for the pre-closure phase of operations, Part 20 was not considered adequate for the post-closure phase as the manner, the timing, and the nature of potential radioactive releases for specific types of LLW would be more difficult to predict under any scenario – natural or otherwise – for any particular repository design. To determine what specific technical requirements might be needed to achieve safety during the post-closure phase, a more definitive assessment of the potential radiological hazard was needed. To conduct this assessment, two logical exposure scenarios came to mind: (a) an event in which radioactive material is transported off-site (i.e., ground water migration) as a result of the natural evolution of the disposal system and its environs – what is now commonly referred to as undisturbed or "base case" scenario"; or (b) a potential event similar to the one already described above in which individuals come into unintentional, direct contact with the buried waste – what is now commonly referred to as a disturbed or "human intrusion" scenario.

The remaining three performance goals were intended to address potential long-term exposures that might be encountered during the post-closure phase (period) of the disposal facility life-cycle. In proposing these performance standards (and the supporting technical requirements), the Commission recognized that the period of greatest reliance on the disposal system would be well-after the facility had been closed as some LLW could still remain hazardous for up to 500 years. Because of the potential for humans to egress into a disposal facility and come into contact with radioactive waste, albeit inadvertently, it was quickly recognized that the intruder scenario would likely be the key scenario driving decisions on what combination of siting and design technical requirements were necessary to provide a level of

---

[28]*ICRP Publication 30* was issued in four parts between 1979 and 1988. See ICRP (1979, 1980, 1981, and 1988). Including indexes and supplements, there are 8 volumes associated with the *Publication 30* series.

protection sufficient to ensure safety of the public and the environment taking into account the hazard posed by certain types of LLW over the long time-frames of concern.[29]

6.3   10 CFR Part 61 Scoping Activities

Based on the review of the past performance of some LLW disposal sites, it was recognized that certain LLW management practices (i.e., siting and design decisions, preferred waste forms, packaging techniques) had already produced favorable disposal outcomes. For some sites, LLW had been contained in disposal cells and no releases of radioactive material to the accessible environment had occurred. The challenge in developing the new regulation was to understand what combination(s) of practices and/or standards could be relied on to produce the same favorable outcomes at future disposal sites. Understanding the answers to theses questions would help address the question as to what level of protection was necessary for a LLW disposal facility.

6.3.1   NUREG-0456: A Proposed LLW Dose Assessment Model
In developing the technical criteria and standards for SLB, it was recognized that it would be necessary to define the concentrations and quantities of waste acceptable for disposal under a LLW regulation. This meant developing an analytical methodology that allowed the interfaces between key components of a LLW disposal system – i.e, specific siting and design features, performance goals, and source terms – to be defined quantitatively. More specifically, for certain key radionuclides and waste forms, the staff needed to understand what existing LLW management practices and/or disposal methods worked best in containing wastes and limiting doses.

One of the early analyses the staff conducted as part of the draft EIS scoping process was the development of a generic LLW dose assessment methodology. For certain key radionuclides and waste forms, the staff sought to identify (and quantify) an optimal set of model parameters (e.g., disposal practices) that could be used to control doses. Using a consistent set of relatively simple exposure pathways, a deterministic dose assessment methodology was proposed in NUREG-0456 (Adam and Rogers, 1978). It was applied to two reference disposal methods (sites) and a preliminary three-tier LLW classification system. Estimated dose impacts could be compared to dose guidelines developed for the study[30] to determine maximum allowable concentrations (limits) of radionuclides appropriate for each of the proposed waste classification tiers through "what-if" types of analyses.

---

[29]If there were complete assurance that a commercial LLW disposal site would not be subject to human intrusion, then the Part 61 rulemaking effort would have been reduced to determining what technical criteria where necessary to ensure that the disposed wastes would remain within the confines of the disposal facility until such time that the LLW had decayed to natural background levels. However, because complete assurance was not possible, the rulemaking effort needed to account for the eventuality that there would be human egress into a disposal site and exhumation of or contact with the wastes, specific design precautions and/or waste form specifications might be necessary to protect against the more hazardous, longer-lived LLW forms, specifically Classes-B and -C wastes (NRC, 1982; 47 FR 57451).

[30]By law, EPA had the responsibility for the development of radiation exposure standards and criteria to be applied to LLW. However, at the time of the staff's scoping analyses, such criteria were not available. Consequently, the NRC staff postulated a reasonable set of guidelines to provide protection from the effects of ionizing radiation, based on a review of the recommendations of national and international standard-setting organizations, consistent with *ICRP Publication 26* (ICRP, 1977). See Adam and Rogers (1978, pp. 6–10). Also see discussion in Section 6.4.1 of this paper for more information.

The dose methodology developed was the traditional release-transport-exposure-consequence model. The methodology consisted of a basic dose model, dose guidelines, exposure scenarios, and calculational basis. Two mechanisms or exposure scenarios were considered in which individuals come into contact with the waste. They were the "on-site reclaimer" scenario and "off-site transporter" scenario. The on-site reclaimer scenario considered six potential exposure pathways versus four for the on-site transporter scenario (see Table 8). These exposure scenarios were considered to be reasonably conservative. All features of the NUREG-0456 dose methodology were deterministic. The intruder scenario was assumed to occur with a probability of one, 150 years after the end of administrative controls at the disposal site, when most of the short-lived radionuclides have already decayed. The off-site transporter scenarios were also calculated deterministically and were assumed to occur immediately after the waste was disposed. In this latter scenario, there is essentially no credit for radionuclide decay and the releases therefore could be considered instantaneous exposures.

Once developed, the overall methodology was benchmarked against existing analog sites to validate the computational features of the analysis. The analog locations selected were the Maxey Flats LLW disposal site, in Kentucky, and the Latty Avenue uranium mill tailings site, in Missouri. In addition, based on the study's dose limit guidelines limits, the methodology was also able to provide preliminary estimates the maximum concentrations or inventories of radioactive material in LLW that were permissible to ensure that dose exposures did not exceed the assumed safety goals for maximum individual and total population dose. Before publication, all features of the NUREG-0456 methodology and results were peer-reviewed to provide a critical, independent assessment of the work.

6.3.2   NUREG/CR-1005: A Proposed Radioactive Waste Classification System

Having defined a generic methodology for understanding the sensitivity (coupling) between key disposal system interfaces, the next phase of the EIS scoping analysis was to propose a waste classification system that would allow a correlation between the hazard posed by the waste, the safety goal to be achieved by disposal, and some prescriptive regulatory requirements necessary to achieve the safety goals. In considering any disposal solution, radiotoxicity coupled with environmental mobility are recognized as key parameters in defining the magnitude of exposure hazards to the public.

For example, analyses from NUREG-0456 already indicated that certain disposal practices such as increasing the time frame when the first exposure occurs through the use of institutional (administrative) controls could limit the magnitude of those exposures or obviate the significance of certain exposure scenarios all together. Alternatively, burying wastes at greater depths can achieve similar dose outcomes by eliminating the potential for certain types of intruder scenarios as well as providing some shielding of the wastes. Hence, by focusing on the length of institutional controls and limiting the physical accessibility of the wastes, it was possible to formulate disposal categories that indicated how specific types of waste should be treated as well as recommend radionuclide concentration limits for each disposal category.

Thus, in considering the importance of half-life (decay) and environmental mobility to potential dose outcomes, King and Cohen (1977) suggested that any one of the following three disposal actions could occur:

- The radioactive waste did not pose significant radiological health-risk to the public, and the waste could be disposed as part of the municipal waste stream.

Table 8. Exposure Scenarios Considered in NUREG-0456. Taken from Adam and Rogers (1978, pp. 15–17).

Scenario	Event	Pathway	Comments
On-Site Reclaimer	Inhalation	Worker	Inhalation of contaminated dust.
		Resident	
	Ingestion	Well Water Consumption	Ingestion of contaminated ground water and/or consumption of food grown in soil irrigated with contaminated ground water
		Food Consumption	
	Direct Exposure	Worker	Direct exposure to gamma radiation.
		Resident	
Off-Site Transport	Inhalation	Continuous Operational Release	Atmospheric transport.
		Accidental Release	
	Ingestion	Groundwater to River	Ingestion from contaminated ground water resource.
		Surface Erosion	

- The radioactive waste did pose some level of radiological health-risk to the public, and the waste needed to be confined in some controlled manner to allow limited releases to the environment at predictably low rates, consistent with levels of natural background of radiation found in nature.

- The radioactive waste posed a significant radiological health-risk to the public, over an extended period of time, and the waste needed to be isolated so that biologically significant releases of radioactive material to the environment (or inadvertent human intrusion) were unlikely.

Thus, for the purposes of scoping the Part 61 regulation, a "what type of radioactive waste goes where" type of disposal classification system was proposed in NUREG/CR-1005 (Rogers, 1979). Five types of disposal solutions were proposed applicable to all types radioactive waste. The principal considerations in defining the proposed disposal categories were the duration of institutional controls and reclaimer accessibility (Op cit., p. 24). As previously noted, it was believed that governmental institutions could restrict access to disposal sites and thus the potential for coming into contact with hazardous wastes if an institutional control time of sufficient duration was specified. If this time were sufficiently long, then the exposure (hazard) would be reduced by virtue of the natural decay of the waste. Similarly, if wastes were buried deeply enough, the same benefit could be achieved by virtue of the isolation of the wastes at depths greater than those reached by routine excavation. Both considerations were key to NRC's Part 61 umbrella regulation concept.

Building upon the earlier work of NUREG-0456, deterministic *disposal concentration guides* (DCGs)[31] applicable to each disposal class were proposed consistent once again with some specified safety goal. DCGs were the front-end parameters of the dose assessment model. They represent the activity of the waste available for consideration in the assessment at the time of disposal. DCGs were derived by starting with a specified dose limit and working backwards through the dose model, pathway-by-pathway, to the initiation point of the analysis. Another important interface value was the *maximum average concentration* (MAC). It represents the back-end of the dose assessment model. It corresponds to the radionuclide contaminant concentration found in a particular exposure pathway. Both concentration parameters are expressed in units of microcuries per cubic centimeter ($\mu Ci/cm^3$). By using a revised dose assessment model (Rogers and others, 1979[32]), it was demonstrated that DCGs and MACs could be used to derive a five-tier system of disposal recommendations taking into account duration of institutional control and reclaimer exposure pathways[33] (see Table 9). In general, the NUREG/CR-1005 analysis indicated that for higher calculated DCGs (see Table 10), additional administrative and isolation measures were needed to ensure the safe disposal of the waste. The analysis also showed that for some exposure scenarios, the MAC can be the limiting factor in the specification of a radionuclide-specific DCG. The analysis also showed

---

[31]Because it is not practical to perform a radioisotopic survey for every type of LLW configuration, DCFs for individual isotopes were developed for NUREG/CR-1005.

[32]Similar to or derived from the NUREG-0456 dose model.

[33]For the purposes of NUREG/CR-1005, only four dose pathways were considered: reclaimer dust inhalation, food ingestion from reclaimed soil, well water consumption, and direct gamma radiation. It is believed that these pathways are the most restrictive in limiting doses to receptors.

Table 9. Waste Disposal Classification Categories Proposed in NUREG/CR-1005. Taken from Rogers (1979, p. 25).

Disposal Class	Administrative Control	Accessibility	Comments
A	None	No reclaimer access	Default class. No upper-limit for DCGs. Understood to be deep geologic isolation.
B	150 years	No reclaimer access except well water after 150 years	Ready access to reclaimer is unlikely. Understood to represent intermediate-depth land burial (about 30 ft). Off-site-transport, well-water ingestion is controlling exposure scenario after 150 years. DCGs are from Class C and adjusted using a 150-year decay factor.
C	20 years	No reclaimer access except well water after 20 years	Ready access to reclaimer is unlikely. Understood to represent intermediate-depth land burial. Off-site-transport, well-water ingestion is controlling exposure scenario after 20 years.
D	150 years	Reclaimer access following administrative control	Understood to represent shallow land burial (about 10 ft).
E	None	Worker/reclaimer access	Understood to correspond to a municipal sanitary landfill.

135

Table 10. DCGs for Waste Classes Proposed in NUREG/CR-1005. Waste classes are defined in Table 9. Taken from Rogers (1979, p. xiv).

Radionuclide	Waste Class DCGs (in µCi / cm³)				
	A*	B	C	D	E
^{3}H	2.9e9	4.3e5	94	94	0.05
^{14}C	7.1e6	140	140	2.4e-3	1.2e-3
^{55}Fe	1.9e10	SA	SA	SA	12
^{66}Co	9.7e9	SA	SA	2.1e6	2.5e-4
^{90}Sr	3.6e8	38	2.4	0.02	2.3e-4
^{99}Tc	1e4	64	64	0.1	0.05
^{129}I	850	0.3	0.3	0.3	0.024
^{135}Cs	2.4e3	20	20	0.2	0.10
^{137}Cs	1.7e8	SA	SA	0.9	4.2e-3
^{235}U	41	11	11	0.03	0.015
^{238}U	6.4	SA	SA	0.03	0.015
^{237}Np	1.3e4	0.3	0.3	0.02	5.4e-4
^{238}Pu	3.4e8	SA	SA	0.4	3.4e-4
^{239}Pu	1.2e6	90	90	0.1	3.0e-4
^{240}Pu	4.7e6	810	810	0.1	3.0e-4
^{241}Pu	2.2e9	SA	SA	5.9e3	0.015
^{242}Pu	7.6e4	13	13	0.1	3.1e-4
^{241}Am	6.4e7	SA	SA	0.4	9.2e-4
^{243}Am	3.6e6	SA	600	0.3	9.2e-4
^{242}Cm	2.6e10	SA	SA	SA	0.024
^{244}Cm	6.2e8	SA	SA	130	1.5e-3

* Specific activity (SA) of the isotope.

that by comparing potential doses with study guidelines, the waste concentrations, waste volumes, or disposal methods could be modified to provide adequate protection to the public – another important EIS scoping consideration.

6.3.3   NUREG-0782: The Draft EIS[34]

The last step in the rulemaking process was to prepare an EIS consistent with NEPA. As noted earlier, the purpose of the draft EIS was to fulfill NRC's NEPA responsibility as well as demonstrate the decision-making process applied to the development of Part 61. Using the EIS process, the NRC staff was able to evaluate the potential health impacts of LLW disposal, possible means for limiting the impacts, and considering these measures, what potential reduction (benefit) could be achieved. The EIS contained an exhaustive and detailed analysis of alternatives such as disposal facility environments, waste characteristics, disposal facility designs, and operating practices. Determinstically-derived doses were presented for whole body and six organs (bone, liver, thyroid, kidney, lung and gastro-intestinal tract. NRC's draft EIS for the proposed Part 61 rulemaking was published as NUREG-0782 (NRC, 1981b). It was a four-volume report prepared following both CEQ regulations for preparing an EIS as well as NRC's NEPA-implementing regulations set out in 10 CFR Part 51 ("Licensing and Regulatory Policy and Procedures for Environmental Protection"). The deterministic analyses in NUREG-0456 and NUREG/CP-1005 provided a generic methodology for evaluating the risks of different types of radioactive wastes and proposing disposal solutions commensurate with the radiological hazard. NUREG-0782 relied on those methodologies and integrated them into the NEPA-EIS framework necessary to support the proposed rulemaking. However, unlike the earlier analyses, NUREG-0782 considered data (information) viewed to be more representative of the types and kinds of LLW being managed at the time as well as pervasive LLW management practices. The NEPA-required analyses were mostly described in Volume 2 of NUREG-0782; Volume I was a "summary" report. Volumes 3 and 4 of the draft EIS contained the technical analyses and other supporting information that addressed the required elements of an EIS.[35]

---

[34]In conducting the draft EIS scoping calculations, the staff assumed a reference disposal facility representative of existing LLW disposal facility designs and operating and management practices throughout the United States. To summarize, it was decided that the draft EIS reference design would be a SLB facility located in a humid environment characteristic of the eastern United States. The reference facility covered an area of 148 acres with a capacity of one million ft³. This general location was selected because that part of the country was determined to be generating most of the LLW and thus most likely to have the largest number of disposal facilities in the future. The site had four distinct climate seasons although the winters were considered short and mild with an average annual precipitation of about 46 inches. The disposal facility was assumed to have a 20- to 40-year operational life cycle with a disposal capacity of 1 million cubic meters. At the end of operations, the disposal site would be stabilized using existing conservation practices, and the site closed and decommissioned. Following decommissioning, the NRC operating license would then be terminated, and title of the site would be transferred to a governmental agency that would provide active institutional controls (surveillance, monitoring, and custodial maintenance) for a period of about 100 years. During this 100-year caretaker period, there would be no incidents involving inadvertent human intrusion.

The reference facility and other applicable parameters are described in more detail in Appendix E ("Description of a Reference Disposal Facility") of Volume 2 of NUREG-0782.

[35] Specifically, the purposes, scope, and need for the rulemaking action, description of the affected environment, discussion of a preferred action as well as consideration of alternatives, costs, and impacts.

Table 11. Waste Steams Considered in Part 61 EIS Scoping Analyses. Taken from Volume 2 of NRC (1981b, p. 3-11).

Waste Stream Group	Group Description
Group I: LWR Process Wastes	PWR Ion Exchange Resins PWR Concentrated Liquids PWR Filter Sludges PWR Filer Cartridges BWR Icon Exchange Resins BWR Concentrated Liquids BWR Filter Sludge
Group II: Trash	PWR Compactible Trash PWR Noncompactible Trash BWR Compactible Trash BWR Noncompactible Trash Fuel Fabrication Compactible Trash Fuel Fabrication Noncompactible Trash Institutional Trash (large facilities) Institutional Trash (small facilities) Industrial SS Trash (large facilities) Industrial SS Trash (small facilities) Industrial Low Trash (large facilities) Industrial Low Trash (small facilities)
Group III: Low Specific Activity Wastes	Fuel Fabrication Process Wastes UF_6 Process Wastes Institutional LSV Waste (large facilities) Institutional LSV Waste (small facilities) Institutional Liquid Waste (large facilities) Institutional Liquid Waste (small facilities) Institutional Biowaste (large facilities) Institutional Biowaste (small facilities) Institutional SS Waste Institutional Low-Activity Waste
Group IV: Special Wastes	LWR Nonfuel Reactor Components LWR Decontamination Resins Waste from Isotope Production Facilities Tritium Production Waste Accelerator Targets Sealed Sources Industrial High-Activity Waste
Abbreviations: LWR	light-water reactor
PWR	pressurized water reactor
BWR	boiling water reactor
SS	sources and special nuclear material
LSV	liquid scintillation vial

Table 12.   Radionulides Considered in Part 61 EIS Scoping Analyses.   Taken from Volume 2 of
NRC (1981b, p. 3-12).

Radionuclide	Half Life (years)	Radiation Emitted	Principal Means of Production
^{3}H	12.3	β	Fission; ^{6}Li (n, α)
^{14}C	5730	β	^{14}Ni (n, p)
^{55}Fe	2.60	γ	^{54}Fe (n, γ)
^{66}Co	5.26	β, γ	^{59}Co (n, γ)
^{59}Ni	80,000	γ	^{58}Ni (n, γ)
^{63}Ni	92	β	^{62}Ni (n, γ)
^{90}Sr	28.1	β	Fission
^{94}Nb	20,000	β, γ	^{93}Nb (n, γ)
^{99}Tc	2.12×10^5	β	Fission; ^{98}Mo (n,γ), ^{99}Mo (β^-)
^{129}I	1.17×10^7	β, γ	Fission
^{135}Cs	3.0×10^6	β	Fission; daughter ^{135}Xe
^{137}Cs	30.0	β, γ	Fission
^{235}U	7.1×10^8	α, γ	Natural
^{238}U	4.51×10^9	α, γ	Natural
^{237}Np	2.14×10^6	α, γ	^{238}U (n, 2n), ^{237}U (β^-)
^{239}Pu	86.4	α, γ	^{237}Np (n, γ), ^{238}Np β^-); daughter ^{242}Cm
^{239}Pu	24,400	α, γ	^{238}U (n, γ), ^{238}U (β^-), ^{239}Np (β^-)
^{240}Pu	6580	α, γ	Multiple n-capture
^{241}Pu	13.2	β, γ	Multiple n-capture
^{242}Pu	2.79×10^5	α	Multiple n-capture; daughter ^{242}Am
^{241}Am	458	α, γ	Daughter ^{241}Pu
^{243}Am	7950	α, γ	Multiple n-capture
^{242}Cm	32	α, γ	Multiple n-capture
^{244}Cm	17.6	α, γ	Multiple n-capture

Table 13. Waste Spectra Considered in Part 61 EIS Scoping Analyses. Each waste spectrum represents a cross section of all waste streams that might be generated and disposed of in a LLW disposal facility. Adopted from Volume 2 of NRC (1981b, p. 3-21).

Waste Spectrum	Description
1	This spectrum assumes a continuation of existing and some past waste management practices. Some of the LWR wastes are solidified. However, no processing done on organics, combustible wastes, or streams containing chelating agents. LWR resins and filter sludges are assumed to be disposal sites in a dewatered form. LWR concentrated liquids are assumed to be concentrated in accordance with current practices, and are solidified with various media designated as solidification scenario A. No special effort is made to compact trash. Institutional waste streams are shipped to disposal sites after they are packaged with currently utilized absorbent materials. Resins from LWR decontamination operations are solidified in a medium with highly improved characteristics.
2	This spectrum assumes that LWR process wastes are solidified using improved solidification techniques. LWR concentrated liquids are additionally reduced in volume through an evaporator/crystallized. All LWR concentrated liquids are evaporated in 50 weight percent solids, and all LWR process wastes are solidified. In the case of cartridge filters, the solidification agent fills voids in the packaged waste but does not increase the volume. Liquid scintillation vials are crushed at large facilities and packed in absorbent material. All compactible trash streams are compacted; some streams are compacted at the source of generation; and some waste streams are compacted at the disposal facility. Liquids from medical isotope production are solidified.
3	In this spectrum, LWR process wastes are solidified assuming that further improved waste solidification agents are used. LWR concentrated liquids are first evaporated to 50 percent weight solids. All possible incineration of combustible material (except LWR process wastes) is performed; some incineration is done at the source of generation (fuel cycle trash, LWR decontamination resins, institutional wastes from large facilities and industrial trash from large facilities) and some at the disposal site (institutional and industrial trash from small facilities). All incineration ash is solidified.
4	This spectrum assumes extreme volume reduction. All wastes amenable to evaporation or incineration with fluidized bed technology are calcined and solidified; LWR process wastes, except cartridge filters, are calcined in addition to the streams incinerated in Spectrum 3. All noncompactible wastes are reduced in volume at the disposal site or at a central processing facility using a large hydraulic press. This spectrum represents the maximum volume reduction that can be practically achieved.

- considering alternative waste form and/or waste packaging configurations;

- evaluating the effectiveness of the duration of institutional controls; and

- examining the effects of engineered and/or natural barriers to intrusion.

Consistent with the draft EIS scoping process and rulemaking objectives, the intent was to understand what effect, if any, potential mitigation actions would have on predicted dose outcomes based on the likely waste streams expected. If effective, the potential mitigation actions might make reasonable regulatory recommendations to advance for the purposes of the Part 61 rulemaking. These recommendations would be decided on by identifying which exposure scenarios were the most restrictive (i.e., producing the highest doses) and then evaluating the effectiveness of the potential mitigation actions listed above in reducing the estimated doses.

Concentration-limited exposure events.[37] There were three potential exposure scenarios identified depending on the time duration (length) of the exposure scenario. The first was a intruder-construction scenario." It assumed that the intruder is unaware of the radiological hazard and constructs of a house directly over a LLW disposal cell. The second scenario was a "intruder-discovery scenario" in which an intruder exhumes a portion of the disposal cell as part of building a house and comes into direct contact with the waste. Realizing that something is amiss, the intruder then terminates construction activities and egresses from the disposal facility but not before receiving a minimal exposure to the ionizing radiation. The third scenario was a "intruder-agriculture scenario" in which intruders grow and consume food on contaminated soil at the site. This was a type of resident farmer scenario. This scenario also assumes the intruder is unaware of the radiological hazard. In all cases, the overriding assumption is that any one of the intrusion scenarios takes place after the end of the governmental 100-year caretaker period, when institutional controls end and knowledge about the LLW facility ceases. In all cases, it was also assumed that the probability of intrusion at year100-post closure was one and that the probability of intruding into Class-C waste was one. Further in all cases the conditions of intrusion, lack of recognition and routes exposure to the intruder can be best characterized as extreme bounding conditions.

Activity-limited exposure events.[38] This scenario occurs when an intruder drills a water-supply well at or in the vicinity of the disposal site, unaware of the presence of LLW, and consumes contaminated water. Four receptor-well locations (scenarios) were considered, as noted below:

---

[37] The EIS' treatment of these three exposure scenarios is discussed in more detail in Chapter 4 ("Presentation and Analysis of Alternatives – Intruder") of Volume 2 of NUREG-0782.

[38] Analysis of these four scenarios is discussed in more detail in Chapter 5 ("Long-Term Environmental Protection – Presentation and Analysis of Alternatives") of Volume 2 of NUREG-0782.

The water supply well...	The exposure scenario involves ...
... penetrates disposal facility cell	...only a single intruder. Would be considered a subsistence well today.
... is located at the edge of the disposal site boundary	... locating a production well for a few individuals in the path of the contaminant plume.
... is located 500 meters down-gradient from the disposal facility	... locating a production well for a small population of individuals (about 100) in the path of the contaminant plume.
... is located 1 km down-gradient from the disposal facility	... locating a production well for a small population of individuals (about 300) in the path of the contaminant plume.

Again, as was the case with the concentration-limited exposure events, there was no knowledge of the radiological hazard posed by the drinking water from the presence of the LLW disposal facility because the scenario takes place after the end of the caretaker period and all institutional knowledge of the facility had been lost.

6.4   Assumed Definition of Safety

As noted earlier, EPA was responsible for developing and issuing environmental standards, guidelines, and criteria to ensure that the public and the environment were adequately protected from potential radiation impacts. President Richard M. Nixon announced the creation of EPA with the publication of *Reorganization Plan No. 3 of 1970*. The intent of this plan was to consolidate Federal research, monitoring, standard-setting and enforcement activities into one agency to ensure environmental protection. This plan also granted EPA its standard setting authority to establish "... generally applicable environmental standards for the protection of the general environment from radioactive material .... (The White House, 35 FR 15624) "

6.4.1   EPA Efforts to Promulgate LLW Standards
In 1972, the EPA Office of Radiation Programs began a program with the Conference of Radiation Control Program Director's to examine the practice of shallow-land disposal of commercial LLW (EPA, 1988; p. 1-3). From 1977–78, EPA conducted a series of public workshops to examine the policy and technical issues associated with the development of radiation standards (e.g., EPA, 1978a). As a precursor to the required standards, and about the same time the NRC was developing a LLW regulatory framework, EPA proposed *Federal Guidance* for the storage and disposal of all forms of radioactive waste (EPA, 1978c; 43 FR 53262). In 1980, Meyer (p. 10) described the sources EPA was consulting in the development of its proposed standards. They included the NEPA statutes, the Biological Effects of Ionizing Radiation (BEIR) II recommendations (NAS, 1977a), *ICRP Publication 26* (ICRP, 1977), and two other reports (NAS, 1977b; and EPA, 1978a). However, the agency later withdrew its proposed guidance criteria noting that the many types of radioactive wastes and different methods necessary to manage and dispose of them made the issuance of generic disposal guidance too complex and that radiation standards based on waste type would be the best approach (EPA, 1981; 46 FR 17567). Alternatively, EPA decided to promulgate regulations specific to the management and disposal of LLW.

In December 1982, the Commission issued the final Part 61 regulation. Following its release, and depending on its final content, the staff noted its intent to amend Part 61 (and potentially other NRC regulations) once the EPA LLW standards were issued if the regulations did not comply with the standards (NRC, 1989c; p. 11). In August 1983, EPA published an ANPR announcing its plans for establishing general environmental radiation protection standards for LLW (NRC, 1983b; 48 FR 39563). In connection with the development of these standards, tentatively designated as 40 CFR Part 193, EPA developed the PRESTO-EPA[39] computer code (EPA, 1983a). Similar to NRC's earlier dose-modeling efforts in this regard, the purpose of the EPA-sponsored code was also to model radionuclide transport through major environmental pathways to humans. EPA also requested that the Agency's Science Advisory Board review a PRESTO-EPA-derived risk assessment prepared as part of the LLW standards development (EPA, 1985b). As a result of these efforts, EPA transmitted a proposed regulation to the Office of Management and Budget (OMB) in 1987. This was followed by the publication of a draft EIS in 1988.

In describing the proposed LLW standards, Gruhlke and others (1989; p. 273) notes that EPA proposed the following definition of commercial LLW:[40]

> "... radioactive waste that was not (1) spent fuel, high-level radioactive waste, or transuranic waste, as previously defined in 40 CFR Part 191, (2) or uranium or thorium mill tailings subject to 40 CFR Part 192, or (3) or NARM as defined in 40 CFR Part 764 ...."

EPA's proposed LLW regulation never cleared the OMB review process. The rule encountered significant interagency opposition during the review because of concerns over the ground-water provisions of the proposed standard (EPA, 2000a; p. 21). Consistent with other regulatory authorities, EPA did successfully promulgate regulations in other nuclear waste management areas – uranium and mill tailings (EPA, 1983c) and HLW (EPA, 1985a). EPA also promulgated standards related to the maximum concentration limits (MCLs) of radioactive material in its *National Primary Drinking Water Standard* found at 40 CFR Part 141 (EPA, 1976)[41] as well as radiation emission standards associated with the Clean Air Act of 1977 (CAA), as amended.[42]

---

[39]PRESTO is an acronym for Protection of Radiation Effects from Shallow Trench Operations. The computer code was a simple one-dimensional ground-water transport model (EPA, 1988; p. 8-2) and although it could be used to estimate intruder exposures, EPA expressed the view in its draft EIS that the intruder pathway was probabilistic in nature and that safeguards against it should be considered on a site-specific basis. For this reason, EPA did not consider the human intrusion scenario it its draft EIS.

[40]A more detailed discussion of the history and evolution of the LLW definition can be found in Appendix C of this paper.

[41]EPA first promulgated interim regulations in 1976 that established MCLs for radium-226 and radium-228 of 5 pCi/L. The most recent MCLs can be found in EPA (2000b, 65 FR 76708), which also includes an MCL of 30 μg/L for uranium.

[42]CAA provided EPA the specific authority to limit radionuclide emissions to the air. Section 122 of the act directed EPA to review all relevant information and determine whether emissions of radioactive pollutants will cause or contribute to air pollution that may reasonably by anticipated to endanger public health.

In 1979, EPA added radionuclides to the list of hazardous air pollutants (HAPS) under the CAA (EPA, 1979). Among

Two key features of the analytical framework used to evaluate the performance of a hypothetical LLW disposal facility in the draft EIS are discussed below. They are *the LLW streams* and *the exposure pathways considered*.

6.3.3.1   The LLW Waste Streams Considered[36]

At the time the regulation was being developed, there were an estimated 20,000 NRC materials licensees producing LLW in a wide variety of waste types, forms, and amounts. LLW was not a uniform physical quantity. It contained both short-lived and long-lived radionuclides. LLW also ranged from trash that was only suspected of being contaminated to highly radioactive material such as activated structural components from nuclear power reactors. It could be in solid, liquid, or gaseous forms.

For the purposes of the draft EIS scoping process and analyses, existing commercial LLW was separated into 36 distinct waste streams (see Table 11). Each waste stream represented a separate type of LLW generated by a particular type of waste source, and had distinct physical, chemical, radiological, and other characteristics unique to that waste stream. The isotopic content of various waste streams was also analyzed. The most important radionuclides present in each commercial waste stream were then identified (see Table 12) for consideration in the draft EIS analysis. To allow for the required consideration of disposal impacts and alternative management options, the volumes of each waste stream were also considered. In developing the regulation, the Commission noted that a key concern was the mobility of certain long-lived radioisotopes (^{129}I, ^{99}Tc, ^{14}C, and tritium) in the environment, especially in ground water. By defining radionuclide concentration limits for each disposal site, the Commission sought to ensure that the proposed Part 61 performance objectives related to ground water would be me (NRC, 1982b; 47 FR 57455).

As noted earlier, another of the Commission objectives in developing the LLW regulation was to identify existing as well as new LLW management practices and designs that could contribute to meeting the overall performance goals. Consequently, it was decided to also consider waste form and processing options as part of the draft EIS scoping analysis. This was achieved by categorizing existing commercial LLW into four waste "spectra" representing generic processing options to be considered (see Table 13).

6.3.3.2   The Exposure Pathways Considered

Based on a review of exposure pathways considered in earlier staff investigations, the NRC selected a limited number of exposure pathways for the draft EIS. There were concentration-limited exposure events and activity-limited exposure events. As noted earlier, all of the intruder scenarios described were assumed to occur with a probability of 1. For each of the scenarios studied, the staff addressed the following four potential mitigation actions in the context of the draft EIS:

- controlling the concentrations of the radionuclides in the specific waste streams being disposed;

---

[36]The waste stream definition process described above is explained in more detail in Appendix D ("Low-Level Waste Sources and Processing Options") of Volume 2 of NUREG-0782.

6.4.2   The NRC Selection of a Default LLW Standard

As noted above, EPA's LLW standards and criteria were not available at the time NRC was developing its LLW regulatory framework. Rather than delay the development of its disposal regulations, the NRC staff decided to postulate a reasonable set of "study guidelines" that could be used as surrogates for the forthcoming EPA standard. At the time, there was no nationally accepted set of safety guidelines defining what level of safety (protection) disposal facilities should provide the public from the ionizing effects of radiation. Consequently, the staff decided to review the literature[43] and consider the recommendations of national and international standard-setting organizations to identify surrogate dose guidelines for the scoping analyses and later, the proposed and final rule. See Table 14.

Then, as now, the ICRP was considered to be the authoritative body on the subject of health physics. In proposing limits on radiation risks, the ICRP observed that radiation risks where a very minor fraction of the total number of environmental hazards to which members of the public where generally exposed. Consequently, in considering what the acceptable magnitude of radiation risk might be, the ICRP suggested that such risks be considered in light of the public acceptance of other risks encountered in everyday life – generally in the range of 10^{-6} to 10^{-5} per year. In its *Publication 26* (ICRP, 1977; p. 23), the ICRP recommended a whole body-dose equivalent of 500 mrem/yr for individual members of a critical group (i.e., tens of individuals) provided that the average annual dose equivalent to individual members of the public (i.e., hundreds of individuals) did not exceed 100 mrem/yr (Op cit.).

To conduct the hypothetical series of dose analyses described in NUREG-0456 and NUREG/CP-1005, the ICRP 1977 recommendations were used as dose guidelines (e.g., surrogate standards). See Adam and Rogers (1978, p. 7) and Rogers (1979, p. 9), respectively. Estimated exposures to workers and the (hypothetical) inadvertent intruder were not given separate treatments in these analyses.

In developing the Part 61 draft EIS (NUREG-0782 – NRC, 1981b), the staff decided to rely on existing EPA standards in allied areas of radiation management and selected a range of public exposure limits from those standards which were expected to bound EPA's forthcoming rule. One mrem/yr was selected as a lower dose bound since at the time it was less than the 4

---

the radionuclides included were those defined by the AEA as source material, special nuclear material, and by-product materials as well as TENORM. EPA determined that radionuclides are a known cause of human cancer and genetic damage, and that radionuclides cause of contribute to air pollution within the meaning of Section 122(a) of the CAA. Once listed, Section 112(b)(1)(B) of the CAA requires EPA to establish "National Emission Standards for Hazardous Air Pollutants" (NESHAPS) at a level which provides an ample margin of safety. In 1989, EPA published NESHAPs for eight radionuclide source categories, covering an estimated 6300 sources at 40 CFR Part 61 (EPA, 1989). Eleven parties, primarily representing the regulated community, subsequently sued EPA during the development of the radionuclide NESHAPs.

Between 1992 and 1996, EPA evaluated the ALARA programs at many NRC-licensed facilities. Based on this evaluation, EPA concluded that radionuclide emissions from NRC- and Agreement State-licensees did not exceed the 10 mrem/yr NESHAP-established standard. NRC subsequently issued a "constraint rule" under Part 20 of its regulations that required licensees to maintain emissions below the 10 mrem/yr standard. EPA found that NRC's regulatory program protects the public health to a safe level with an ample margin of safety and the NESHAP regulating air emissions from NRC-licensees was rescinded in 1996 (EPA, 1996).

[43]See Appendix N ("Analysis of Existing Recommendations, Regulations, and Guides") in Volume 4 of NUREG-0782.

Table 14. Dose Guideline Options Considered by the NRC in Developing 10 CFR Part 61. Taken from the references cited.

Receptor	NUREG-0456 (Adam and Rogers, 1978)	NUREG/CP-1005 (Rogers, 1979)	NUREG-0782 (NRC, 1981)	Draft Part 61 (NRC, 1981)	Final Part 61 (NRC, 1982b)
Public (General Population)	Individual exposures to a few individuals (~10s) – 500 mrem/yr [a]	Individual exposures to a few individuals (~10s) – 500 mrem/yr [a]	25 mrem/yr whole-body exposure to an individual at the disposal site boundary [c]	25 mrem/yr whole-body exposure to an individual at the disposal site boundary	25 mrem/yr whole-body exposure to an individual at the disposal site boundary
	Individual exposures to many individuals (~100s) – 100 mrem/yr [a]	Individual exposures to many individuals (~100s) – 100 mrem/yr [a]		Meet EPA requirements of 40 CFR Part 141 for the nearest drinking water supply [e]	
Worker			10 CFR Part 20 [b]	10 CFR Part 20 [b]	10 CFR Part 20 [b]
Intruder			500 mrem/yr [d]	500 mrem/yr [d]	Not specified but implied [f]

a. NUREG-0456 dose guidelines based on recommendations of the ICRP (1977).
b. Includes consideration of ALARA principles.
c. Based on view that releases would not be higher than standards already established for fuel cycle facilities found at Part 190 (EPA, 1977). Commission considered a range of 1 mrem/yr to 25 mrem/yr.
d. Considered to be an unusual event. Dose guidelines in NUREG-0782 and Draft Part 61 based on recommendations of the ICRP (1977).
e. Specifically, maximum radiation concentration limits of 10pCi/l above background levels (or 4mrem/yr whole body exposure). See EPA's *National Primary Drinking Water Standards* (EPA, 1976).
f. Tied to Table 1 concentration limits in draft Part 61 regulation but 500 mrem/yr retained as a basis for limits specified in the tables in the final regulation.
g. Note the technical bases for dose limits under § 61.41; the basis for the concentration limits in the intruder scenario; and the current Part 20 are different. For short lived radionuclides the difference is negligible; for long lived radionuclides the difference may be significant.

146

mrem/yr limit found in the EPA 1976 drinking water standards (EPA, 2000; 65 FR 76710). Bearing in mind that NRC's goal was to propose a LLW regulation based on currently available technology, the staff believed that 1 mrem/yr would provide a limit against which the effectiveness of current technology could be analyzed. At the upper end, 25 mrem/yr was selected as an upper bound since it was already in use as an existing radiation standard at 40 CFR Part 190 – "Environmental Radiation Protection Standards for Nuclear Power Operations" (EPA, 1977) – applied to routine operating releases from nuclear fuel cycle facilities. In proposing this range, the Commission concluded that the forthcoming EPA LLW standards would not be higher than those already set out in Part 190 (NRC, 1981c; 46 FR 38063). The specified performance objective in Part 20 was applied to worker safety because the standard was already being applied to other NRC-licensed facilities and therefore was still considered appropriate to apply to an operating commercial LLW disposal facility. Because the human intruder scenario was considered to be an unusual (rare) event, likely to involve only one or two individuals, the Commission believed that whole body-dose equivalent of 500 mrem/yr (assuming a 100-year period of institutional controls) was considered acceptable and protective, consistent with the earlier recommendations of the ICRP (Op cit.).

From the draft EIS scoping analyses, the staff was able to conclude that a limit in the range of existing EPA drinking water regulations (4 mremyr) could be achieved at the nearest public drinking water supply given some modest increased costs and changes to the reference disposal facility design. The staff also concluded that meeting the EPA drinking water standards at the nearest public drinking water supply would result in annual potential exposures of less than 25 mrem whole body, 75 mrem thyroid, and 25 mrem to any other organ to an individual who might consume water from a well located at the site boundary. An annual exposure limit of 25 mrem whole body, 75 mrem thyroid and 25 mrem to any other organ to the maximally exposed individual at the site boundary coupled with an annual population limit of 4 mrem at the nearest public drinking water supply was, therefore, selected as the preferred performance objective when the regulation was published for public comment (NRC, 1981; 46 FR 38063).

Following a review of the public comments received on its proposed regulation, the NRC made two changes to the final rule as it related to the Subpart C performance objectives. The first was in response to a comment from EPA, which expressed the view that it was inappropriate to apply the agency's drinking water standard in the manner first proposed in §61.41 (NRC, 1982b; 47 FR 57448). The Commission deleted that provision from its final rule. The second comment concerned the proposed 500 mrem whole body dose to the human intruder. Many commenters suggested that the intruder performance objective was too restrictive. It was also argued that a licensee would not be able to monitor or demonstrate compliance with a specific dose limit to an event that might occur sometime in the future several hundred years from now (47 FR 57449). The Commission deleted this provision from the Subpart C performance objectives as well but retained the 500 mrem as a basis for the waste classification limits.

6.4.3   The NRC Proposed LLW Classification System
As a means of relating waste characteristics to the Subpart C performance objectives, a simple waste classification scheme was devised and incorporated into the proposed regulation. This three-tier classification system was based on the earlier thesis demonstrated during the rulemaking scoping process that waste characteristics provide some level of assurance that the performance objectives will be met. Key decision parameters in the waste classification system

were the physical stability of the waste form[44] and its isotopic concentration. These parameters were viewed as important for they provide the minimum information necessary for basic decisions on the safe handling and disposal of wastes.

Three classes of LLW were defined in §61.55 as acceptable for disposal in near-surface facilities. They were designated *Class A*, *Class B*, and *Class C*, with the highest being Class C. Class designations were tied to certain minimum requirements and stability requirements,[45] and specifications for maximum allowable concentrations of certain radionuclides in each class. By controlling isotope concentrations in each waste class (and to a lesser degree, the site inventory), inadvertent intruder exposures are controlled (47 FR 57455). *Class-A waste* includes primarily lightly-contaminated paper, cloth, and plastics. These wastes must be segregated from other LLW during disposal because of their potential for degrading over time and causing subsidence in disposal cells. The isotope concentrations in this class of wastes are not to exceed the values listed in the regulation. *Class-B waste* by definition meets more rigorous physical stability requirements than Class-A wastes. This waste class is also permitted higher isotope concentrations. The physical form and characteristics of Class-B waste must also the meet the *minimum* and *stability* requirements of the regulation. *Class-C waste* was generally considered intruder waste. This higher-activity, longer-lived LLW is generally suitable for SLB and requires special measures to protect against human intrusion after institutional controls lapse. The regulation required that any Class-C waste with concentrations of radionuclides that would cause exposures greater than 500 mrem need to be protected from intrusion by deeper burial and/or through the use of some type of engineered intruder barrier.[46] Wastes exceeding the Class-C concentration limits are, by regulation at §61.55(a)(2)(iv), were "generally not acceptable" for SLB.

As noted in the preceding sections of this paper, the Part 61 regulation is deliberately structured around the three-tier LLW classification system defined by the concentration of radionuclides in the waste form as well as the physical characteristics of the waste form. This classification system is integrated with the stylized human intrusion scenarios that form the basis for the

---

[44]In the *Statements of Consideration* for the final rule (47 FR 57457) , the Commission noted that "...waste that is stable for a long period helps to ensure the long-term stability of the site, eliminating the need for active maintenance after the site is closed. This stability helps to assure against water infiltration caused by failure of the disposal covers and, with the improved leaching properties implicit in a stable waste form, minimizes the potential for radionuclide migration in ground water. Stability also plays an important role in protecting an inadvertent intruder, since the stable waste form is recognizable for a long period of time and minimizes any effects from dispersion of the waste upon intrusion...." The Commission also noted its belief that ".... to the extent practicable, waste forms or containers should be designed to maintain gross physical properties and identity over 300 years, approximately the time required for Class B waste to decay to innocuous levels...." (Op cit.)

[45]The minimum requirements that all waste forms must meet, to be acceptable for near-surface disposal, are given in §61.56(a). In addition to these minimum requirements, certain wastes (i.e., Classes-B and -C wastes, and Class-A waste that is to be co-disposed with Classes-B and -C waste) must be stabilized (structurally) and meet the requirements of §61.56(b). Stability is defined in terms of the ability to keep dimensions and form under disposal conditions. Stability can be provided by the waste form (e.g., activated metals); by processing the waste to an acceptable form (e.g., cement solidification); placing the waste in a high-integrity container (HIC); or by the disposal unit itself (e.g., vault disposal).

[46]The calculation performed to establish the Class-C limits was based on a postulated SLB disposal method. These limits are considered conservative by the Commission since there may be other near-surface disposal methods (and costs) than SLB (NRC, 1987; 52 FR 5999).

Subpart C performance objectives. Despite this rigor, the Commission decided to allow for the consideration of alternative requirements for the classification of LLW at §61.58 on a specific basis so long as it can be demonstrated that the Subpart C performance objectives can be met.

Section 61.58 acknowledges the need to allow for the disposal of different types, physical forms, and quantities of LLW not necessarily recognized at the time the regulation was being developed.

6.4.4 Summary: Final 10 CFR Part 61

The Commission's final LLW disposal regulation at Part 61 was developed with the intent to address some of the past LLW site performance concerns as well as to develop guidelines that could be used to establish technical criteria for selecting, evaluating, licensing, and operating new commercial disposal sites. The regulation covers all phases of shallow, near-surface LLW disposal from site selection through facility design, licensing, operations, closure, post-closure stabilization, to the period when active institutional controls end. Key provisions of the regulation include:

- Specifying minimum geologic/geomorphic characteristics of an acceptable LLW disposal site using the site suitability requirements at §61.50.

- Defining a three-tier waste classification system for commercial LLW disposal based on the concentrations of the longer-lived radionuclides at §61.55.

- Specifying the minimum requirements that all commercial LLW forms must meet at 61.56(a) to be acceptable for near-surface disposal. In addition to these minimum requirements, certain LLW classes[47] must be structurally stabilized and meet the requirements at §61.56(b).

- Introducing requirements for caretaker oversight of LLW disposal sites for a period of 100 years following facility closure at §61.59.

The regulation also establishes procedures, criteria, terms, and conditions on which the Commission would issue and renew licenses for the shallow-land burial of commercially-generated LLW.

In issuing its final regulation, the NRC staff prepared a final EIS, in response to public comments received on the draft EIS and the proposed rule. The final EIS, designated NUREG-0945 (NRC, 1982a), presents the final decision bases and conclusions (costs and impacts) regarding NRC's LLW regulation. In addition, the document refined the deterministic EIS impact analysis methodology and grouped the disposal alternatives into four cases: past LLW disposal practices, existing LLW disposal practices, disposal practices based on proposed final Part 61 regulatory requirements (47 FR 57446), and an upper bound exposure example.

Although the Commission left several of the proposed Part 61 regulations substantially unchanged following the public comment period, the final EIS provided a number of clarifications for specific rule provisions, including the following:

---

[47] Classes B and C, and Class A waste that is to be co-disposed of with Classes B and C waste.

- Doses were generally presented only for the whole body, thyroid, and bone.

- Waste classification represented a combination of waste form, radionuclide characteristics, radionuclide concentration, method of emplacement, and to some extent site characteristics.

- The limits for Class-A and Class-C waste disposal were re-evaluated.

- The Class-C limits were raised by a factor of 10 for all radionuclides.[48]

- A fourth class of LLW – GTCC LLW – was considered generally unacceptable for near-surface, shallow-depth disposal.[49]

---

[48]It should be noted that the concentration limits were established based on the staff's understanding at the time of the characteristics and volumes of LLW that would be reasonably expected to the year 2000, as well as potential disposal methods (52 FR 5999).

[49]In 1986, the NRC staff updated the impacts analysis methodology used in the EIS scoping and rulemaking process to allow for improved consideration of the costs and impacts of treating and disposing of LLW that was close to or exceeding the Class-C concentration limits. See Oztunali and Roles (1986) and Oztunali and others (1986). The updates included the use of the more recent health physics guidance found in *ICRP Publication 30*.

7 The Management of GTCC LLW

Quantities of LLW whose radionuclides concentrations exceeded certain values were defined in §61.55(a)(4)(iv). Such classified wastes are designated as GTCC. They are produced in small volumes primarily as a result of the operation of commercial nuclear power reactors and other fuel cycle facilities. Examples include activated metal hardware (e.g., nuclear power reactor control rods), some spent fuel disassembly hardware (Stellite balls), some ion exchange resins, filters, evaporator residues, some sealed sources that are used in medical and industrial applications, and moisture and density gauges. The radionuclides that are frequently contribute to wastes being classified as GTCC waste include those found in §61.55, Table 2. By law, DOE is responsible for disposing of GTCC wastes.

7.1 NRC Activities

In an 1987 ANPR, the Commission proposed to redefine the existing definition of high-level radioactive waste (HLW) in a manner that would apply the term "high-level radioactive waste" to materials in amounts and concentrations exceeding numerical values that would be stated explicitly in the form of the table (52 FR 5992). The Commission proposed to classify wastes as HLW or non-HLW wastes. Wastes that could not be disposed of safely in a hypothetical "intermediate" disposal facility would be classified as HLW (NRC,1987; 52 FR 5996). The technical basis supporting this proposal was published in Kocher and Croff (1987).

Following a review of public comments on the ANPR, the Commission adopted an alternative strategy. In 1988 (53 FR 17709), the NRC published its proposed amendments to Part 61 recommending, in the first instance, disposal in a separate facility licensed under Part 60 ("Disposal of High-Level Radioactive Waste Geologic Repositories") – the generic regulations for the disposal of HLW (NRC, 1983a). The Commission expressed the view that given the quantities of waste of concern[50] and the likely costs of disposal, a separate disposal facility unique to GTCC LLW was not justified. That same year, the Congressional OTA (1988) published an independent report with its recommendations on the issue that generally supported the Commission's 1988 proposed rulemaking position. Both OTA and the Commission took the position that if, following a review, it was determined that the impact of GTCC LLW disposal on any HLW repository was unacceptable, then DOE should develop an alternative disposal concept. Amendments to Part 61 were proposed that would require the deep geologic disposal of GTCC LLW unless an alternative means of disposal elsewhere was approved by the Commission. This action was proposed to obviate the need for amending the existing classifications of LLW and HLW, thereby insuring that GTCC LLW would be disposed of in a manner consistent with the protection of public health and safety. Following a review of public comments, in 1989, Part 61 was amended at §61.55(b)(2)(iv) to permit the disposal of GTCC LLW in a HLW geologic repository licensed under Part 60 or some other type of disposal facility design approved by the Commission (NRC, 1989a; 54 FR 22578).

On November 2, 1995, the Commission received a petition from the Portland General Electric Company (the utility licensed by the NRC to operate the Trojan Nuclear Power Plant) requesting that NRC's regulations at 10 CFR Part 72 ("Licensing Requirements for the

---

[50]Expected to be in the range of 2000 to 4800 cubic meters through 2030, citing DOE estimates (54 FR 22580). This volume corresponds approximately to a single emplacement drift in a HLW repository.

Independent Storage of Spent Nuclear Fuel and High-Level Radioactive Waste" – at the time) be amended to specifically provide for storage of GTCC waste at an independent spent fuel storage installation (ISFSI) or a monitored retrievable storage (MRS) facility pending its transfer to a permanent disposal facility. See NRC (1996a). Because interim storage of the GTCC waste would be accomplished in a manner similar to that used to store spent fuel at an ISFSI, the petitioner believed public health and safety and environmental protection would be ensured. The NRC staff evaluated the petition and the six comments received during a public comment process,[51] which all supported the petition, and concluded that the petitioner's concept had merit because there are currently no routine disposal options for GTCC waste. The Commission subsequently amended Part 72 to allow licensing for the interim storage of GTCC waste in a manner that is consistent with current licensing for the interim storage of SNF. See NRC (2001). The amendments only applied to GTCC LLW wastes generated at commercial nuclear power plants.

7.2   DOE Activities

The 1988 OTA assessment (p. 31) expressed the view that it would be 15 to 20 years before disposal access for GTCC LLW would be available to generators. As an interim measure, OTA recommended extended on-site storage for those producers who had capacity to do so. For those who had no capacity, OTA recommended storage at an NRC-licensed DOE disposal facility (Op cit.). In 1989, when the issue of the need for the potential for Federal interim storage of nuclear waste was examined, there was no reference to the management of GTCC LLW. See Monitored Retrievable Storage Review Commission (1989).

Section (3)(b)(1)(D) of the LLWPAA directed the Secretary of Energy to issue a report recommending safe disposal options for GTCC LLW. Such a report was issued by the Secretary in 1987. The report (DOE, 1987) also described the types and quantities of GTCC LLW being generated at the time. Hulse (1991) and Lockheed Idaho Technologies Company (1994a and 1994b[52]) have provided updates that later revised earlier information about estimates of current and future volumes of GTCC LLW from the original 1987 census.

DOE published a *Notice of Inquiry* (NOI) in 1995 soliciting public and stakeholder input to the development of a strategy for the management and disposal of GTCC LLW (DOE, 1995; 60 FR 13424). In its *Federal Register Notice*, DOE proposed to prepare a preliminary EIS that indicated its intent to begin the scoping process for developing GTCC LLW disposal options. The scoping process included three public meetings with stakeholders. Five strategy options were proposed in the 1995 NOI. It was noted that the decision-making regarding the

---

[51]The NRC published a notice of receipt of the petition in the *Federal Register* on February 1, 1996 (NRC, 1996a; 61 FR 3619), allowing a 75-day comment period. The NRC received six comment letters, all supporting the petition. The NRC staff evaluated the petition and the comments and concluded that the petitioner's concept had merit. The requirements at Part 72 only provide for licensing storage of spent fuel at an ISFSI and storage of SNF and solid HLW at an MRS. Nonetheless, a reactor licensee could elect to store GTCC LLW at an ISFSI site under licenses issued under other NRC regulations, namely, Part 30 and Part 70. However, the Part 30 and Part 70 regulations at the time did not provide specific licensing criteria for storage of GTCC LLW at an ISFSI, and thus may not have been known to the petitioner or to the commenters that GTCC waste can be stored under a Part 30 or a Part 70 license.

[52]This study concerned sealed sources, the number of which in the United States was estimated to be about 250,000.

Department's preferred management option would be addressed in supplemental NEPA documentation (60 FR 13425). Following the conduct of three public meetings, no additional action was taken by the Department to develop the preliminary EIS. Alternatively, in 2005, the Department published a advance NOI prepare a EIS for GTCC LLW. See DOE (2005, 70 FR 24775).[53] As part of the EIS development process, DOE proposed that the NRC staff participate as a cooperating agency (NRC, 2005). After review, the Commission rejected this proposal and in a 2005 SRM, directed the NRC to comment on the GTCC EIS.

---

[53]It should be noted that in a review of potential waste streams for a HLW repository, another DOE program office has reviewed the characteristics of GTCC wastes. See ORNL (1992). In the final EIS for the Yucca Mountain geologic repository, DOE accounted for GTCC LLW disposal in a bounding analysis that estimates the environmental impacts of repository disposal activities (DOE, 2002; pp. A-57 – A-61). However, there are no published plans at this time suggesting that DOE will place GTCC waste in the proposed HLW repository.

8   OTHER NRC LLW PROGRAM DEVELOPMENTS

Section 5 of this paper describes the regulatory products the staff prepared to help potential licensees develop complete and high-quality license applications based on Part 61 requirements. These products also provide instructions to the staff on how to review those license applications.

In addition to the development of guidance products, the NRC staff has undertaken a number of initiatives intended to aid in the implementation of NRC's LLW regulatory framework. These initiatives are described in Section 7.1 of this paper and took place at various times over the years in relation to the development of the Part 61 regulatory and guidance framework previously described. As part of an agency-wide planning initiative in the early 1990s, the NRC staff undertook a broad reassessment of its LLW program. This reassessment is described in Section 7.2 of this paper.

8.1   LLW Regulatory Guidance and Policy

The NRC staff has historically relied on the use of guidance documents such as technical positions as a means of interpreting the Commission's regulatory requirements. In addition, the Commission periodically issues *Policy Statements* as a means of communicating to licensees and stakeholders Commission views bearing on some particular issue. These "communications" were not intended as substitutes for the regulations and compliance with them is not required. They generally represent the staff's recommendations on preferred approaches to addressing the requirements[54] or the Commission's views on issues bearing on its regulatory activities. Table 15 summarizes the subject areas for which the Commission has issued policy statements or the staff have provided additional regulatory guidance to potential LLW licensees.

The NRC also sponsored numerous technical assistance projects intended to provide predictive models and analytical tools necessary to evaluate the performance of LLW disposal facility systems and components. Areas of interest included waste package container performance, evaluation of leaching phenomena, hydrogeological and hydrochemical characterization and modeling, and cover performance. Most of this work focused on SLB disposal facilities. The use of predictive models to evaluate the performance of a disposal system or its components is generally referred to as "performance assessment" and has gained increased use in NRC's waste management programs (Eisenberg and others, 1999). As early as 1987, the staff recognized that some type of assessment methodology would need to be acquired" or "developed" for estimating the performance of Part 61 LLW disposal facilities (NRC, 1987; 52 FR 5996). To provide focus and integration of the overall LLW program, a LLW performance assessment strategy was also developed (Starmer and others, 1988). A proposed LLW performance assessment methodology (PAM) based on this strategy was subsequently developed by the Sandia National Laboratories (SNL).

---

[54]In general, the staff has believed that methods and solutions differing from those set out in guidance documents should be acceptable if they provide a sufficient basis for the findings requisite to the issuance of a permit or license by the Commission.

Table 15. Additional NRC Technical Guidance and Policy Direction in the Area of LLW.

Title	Scope	Reference
Commission Policy/Position Statements		
"Policy Statement on Low-Level Waste Volume Reduction" [a]	Licensees are encouraged to establish programs to result in good volume reduction practices in order to: (1) extend the operational life of existing commercial LLW disposal sites; (2) alleviate concerns regarding existing LLW disposal capacity should there be delays in establishing regional disposal facilities; and (3) reduce the number of LLW shipments.	NRC (1981c)
"Regulatory Issues in Low-Level Radioactive Waste Performance Assessment" (SECY-96-103) [b]	The Commission expressed its views on: (1) consideration of future site conditions, processes, and events; (2) performance of engineered barriers; (3) specification of a time frame for a LLW performance assessment; (4) treatment of sensitivity and uncertainty in LLW performance assessments; and (5) the role of performance assessment during the operational and closure periods.	NRC (1996)
Technical Positions/Recommendations		
Branch Technical Position on "LLW Burial Ground Site Closure and Stabilization" NUREG-0782	In closing and stabilizing a LLW disposal facility, the overall objective is to leave the site in a condition such that the need for active ongoing maintenance is eliminated, and only passive surveillance and monitoring are required to the point when the NRC license is terminated.	NRC (1979)
Branch Technical Position on "Site Suitability, Selection, and Characterization" (NUREG-0902)	Provides the staff's interpretation of: (1) the site suitability requirements proposed in §61.55; (2) the site selection process as related to the consideration of alternatives, as required by the NEPA process; and (3) the scope of site characterization activities necessary to develop site-specific data necessary for a Part 61 license application and environmental report.	Siefken and others (1982)
Technical Position on "Waste Form"	Provides guidance on acceptable methods for demonstrating compliance with the waste form structural stability requirements found at §61.56.	NRC (1991b)
Branch Technical Position on "Concentration Averaging and Encapsulation"	Defines a subset of concentration averaging and encapsulation practices that the staff would find acceptable in determining the concentrations of §61.55 tabulated radionuclides	NRC (1995a)

Title	Scope	Reference
"A Performance Assessment Methodology for Low-Level Radioactive Waste Disposal Facilities – Recommendations of NRC's Performance Assessment Working Group" (NUREG-1573) [c]	Describes (1) an acceptable approach for systematically integrating site characterization, facility design, and performance modeling into a single performance assessment process; (2) five principal regulatory issues related to the interpretation and implementation of the Part 61 performance objectives and technical requirements, all of which are integral to an LLW performance assessment; and (3) how to implement the NRC's PAM.	NRC (2000)

a. The *Policy Statement* acknowledged but did not specifically identify LLW volume reduction technologies under review at the time. See Trigilio (1981). In a report prepared for the ACNW, Long (1990) examines the use of incineration as a potential volume reduction method.
b. Commission's positions were later restated in NUREG-1573.
c. See Appendix D.

In terms of measuring disposal facility designs and performance against the Part 61 performance objectives, the guidance provided by NUREG-1199, NUREG-1200, and NUREG-1300 was general, and many specific implementation issues and acceptable approaches for resolving them were not addressed. Moreover, the relationships between the overall Part 61 data and design requirements, and the specific LLW performance assessment needs, were not explicitly addressed by the existing guidance documents. Previously, site characterization, facility design, and performance modeling were activities that heretofore were considered separate. To clarify these issues and other issues, the staff developed detailed information and recommendations, for potential applicants, as they relate to the performance objective concerned with the radiological protection of the general public (at §61.41) in NUREG-1573 (NRC, 2000). See Appendix D of this paper for additional information.

8.2 LLW Research

Once NRC's regulatory framework was established, the staff focused its attention to conducting technical analyses intended to provide an improved understanding of the behavior of a LLW disposal facility and its components based on lessons-learned at commercial and DOE-operated LLW disposal sites. Many of the NRC products and activities described elsewhere in this paper where conducted by or on behalf of NRC's Office of Nuclear Materials Safety and Safeguards (NMSS). Another NRC program office, the Office of Nuclear Regulatory Research (RES) also sponsored a significant amount of LLW technical work, or more specifically "research." In 1989, RES staff published a LLW research program plan which presented RES' strategy for pursuing the LLW research studies. See O'Donnell and Lambert (1989). Many of the RES-sponsored research projects completed through 2000 in the area of LLW were cited in NUREG-1573. Appendix E of this report contains a selected bibliography of technical reports and papers sponsored by RES in the LLW area since NUREG-1573 was published.

As noted earlier in this paper, the USGS was given certain basic and applied research responsibilities in the area of LLW. In April 1992, the USGS cooperated with RES on basic research applied to LLW siting, monitoring and modeling issues through an Interagency MOU. A major accomplishment of the joint MOU was the convening of a "Joint USGS–USNRC Workshop on Research Related to Low-Level Radioactive Waste Disposal, May 4–6, 1993, National Center, Reston, Virginia." The workshop covered five topics: (a) surface- and ground-water pathway analysis; (b) ground-water chemistry; (c) infiltration and drainage; (d) vapor-phase transport and volatile radionuclides; and (e) ground-water flow and transport field studies. The workshop and its subsequent proceedings (Stevens and Nicholson, 1996) reported on the current state-of-the-art and practice in research related to LLW disposal hydrogeologic, hydrologic, geochemical and performance assessment issues at commercial and military-related facilities. Presenters and participants were from academia, DOE national laboratories, consulting companies, Federal and State agencies, and international research centers.

Other related workshops on modeling and monitoring and their published proceedings include: (a) NUREG/CP-0163, "Proceedings of the Workshop on Review of Dose Modeling Methods for Demonstration of Compliance with the Radiological Criteria for License Termination" (Nicholson and Parrott (1998); (b) NUREG/CP-0177, "Proceedings of the Environmental Software Systems Compatibility and Linkage Workshop" (Whelan and Nicholson, 2002), which helped to initiate the MOU on multimedia environmental modeling signed by 9 Federal agencies;[55] and (c) NUREG/CP-0187, "Proceedings of the International Workshop on Uncertainty, Sensitivity, and Parameter Estimation for Multimedia Environmental Modeling" (Nicholson and others, 2004). These workshop proceedings highlight the advancements in environmental modeling and performance assessments since the 1993 USGS–NRC LLW workshop which are applicable to LLW issues.

Numerous technical reports and technology transfer workshops have been issued and sponsored by RES. Of particular significance to LLW are NUREG/CR-6805, "A Comprehensive Strategy of Hydrogeologic Modeling and Uncertainty Analysis for Nuclear Facilities and Sites" (Neuman and Wierenga, 2003) and NUREG/CR-6843, "Combined Estimation of Hydrogeologic Conceptual Model and Parameter Uncertainty" (Meyer and others, 2004) which discuss guidance and tools for modeling hydrogeologic systems and radionuclide transport relevant to LLW.

8.3   Strategic Planning

In addition to the guidance development activities described above, in the early 1990s, the staff undertook a broad reassessment of its LLW program taking into account factors outside the control of the NRC. This assessment took place at the time other reviews of the national program were taking place (e.g., GAO, 1992a).

As part of the NRC's first assessment, the staff categorized strategies and options for the Commission to consider to advance the goals and objectives of the LLWPAA. These included: expanding technical assistance, revising the existing Part 61 regulatory framework, seeking greater public involvement in the current LLW program, and passing additional Federal LLW legislation (Taylor, 1993). Focusing on the option to revise Part 61, the staff identified specific areas in the regulation that would make potential candidates for revision with the goal of enhancing public health and safety through the establishment of more precise regulations and addressing the State's experiences in applying the existing Part 61 regulatory framework.

---

[55] See http://www.ISCMEM.org.

Candidate areas identified in the current regulation proposed for revision are listed in Table 16 and include so-called "active" disposal concepts.[56]

At the time these candidate areas were proposed, the staff took the position that there was no evidence that the current regulatory framework represented an impediment to the development of new LLW disposal facilities (Taylor, 1993; p. 6). In fact, it was the staff's view as well as that of several of the Agreement States that major revisions to Part 61, along with the requirement for conforming revisions by the Agreement States, could create instability in current LLW licensing efforts (Op cit., pp. 6–7).

As an alternative to revising specific sections of the regulation, the staff proposed to revise Part 61 by removing its existing specificity, and making it more performance-oriented by placing greater emphasis on the overall performance objectives. This proposal was introduced before the Commission published its *Probabilistic Risk Assessment (PRA) Policy Statement* (See Appendix B of this paper). Under such an approach, the staff would develop guidance documents to address siting, design, construction, operation, closure, and waste form issues.[57] There is no information to suggest that the Commission responded to the staff's 1993 analysis. That analysis was first overtaken in 1995 by the issuance of a Commission Paper – SECY-95-201, entitled "Alternatives to Terminating the NRC's Low-Level Radioactive Waste Disposal Program" (NRC, 1995c) – that described three options regarding the future of NRC's LLW program.[58] In that SECY paper, the staff recommended reducing NRC's LLW program by eliminating or reducing various parts of the program taking into account current developments in the national LLW program as well as reduced budget allocations at the NRC. The ACNW provided it views regarding the staff's recommendations in SECY-95-201 in a letter dated December 29, 1995. See Section 8.3.1 of this paper.

Later, in 1995, SECY-95-201 was overtaken by the Commission's *Strategic Assessment and Rebaselining Imitative*. This was a four-phase strategic planning exercise, the goal of which was to assess and rebaseline NRC's regulatory activities in order to provide a sound foundation for future agency direction and decision-making. The principal focus of the initiative was the identification of key strategic issues associated with NRC's primary responsibility to protect public health and safety and the environment. These key issues were called *Direction-Setting Issues* or DSIs. For each of the 16 DSIs, background papers were developed containing the

---

[56]The staff generally defined active disposal concepts to include retrievablility, active maintenance and monitoring, and a longer period of custodial oversight (Op cit., p. 7).

[57]At the time, the staff estimated that it would take 2 to 3 years to complete a performance-based rulemaking and an additional 3 years for the Agreement States to adopt it.

[58] These options can be briefly described as: (a) continue the program as currently in place; (b) reduce the program by eliminating or reducing various parts; and (c) terminate all parts of the LLW program.

Table 16. Potential Candidate Areas in 10 CFR Part 61 Identified for Amendment by the NRC Staff in 1993. Taken from Attachment B to NRC (1993).

10 CFR Part 61			1983 NRC Staff Recommendation (Attachment B)
Requirement	Subpart	Subject Area	
§61.29	B	Active Maintenance	In conjunction with a longer time period of institutional control, include provisions in the regulation for more inspections and preventive maintenance of the disposal facility following closure to assure that the facility is performing as intended.
§61.41	C	Performance Objectives	Establish more stringent dose requirements for protection of the general population lower than the current 25 mrem/yr.
§61.50	D	Technical Requirements for Land Disposal Facilities	Develop specific technical criteria to cover disposal in above ground vaults (AVGs), which are not currently addressed in the regulations.
§61.50(a)	D	Site Suitability Requirements	Current requirements considered to be "minimum" basic requirements. Past experience indicates more specific siting and design requirements are needed. More credit also needed for performance of engineered barriers to compensate for site deficiencies.
§61.53	D	Environmental Monitoring	In conjunction with a longer time period of institutional control, include provisions in the regulation for a period of environmental monitoring after the 100-year caretaker period.
§61.59(a)	D	Land Ownership	Consideration should be given to assigning a responsible third party to the caretaker role other than the government.
§61.59(b)	D	Institutional Control Period	Extend governmental care taker period for more than 100 years.
§§61.55 and 61.56	D	Waste Classification and Characterization	Specific concentration-averaging requirements are not specified in the regulations
n/a	n/a	Retrievability Option	Currently, there is no provision in the regulation to require that the wastes be recoverable should the disposal facility fail to perform as intended.
n/a	n/a	Ground-Water Protection Requirements	The regulation could be made more explicit on how the ground-water resource would be protected. ACNW has previously recommended specific regulatory action in this area.

Commission's preliminary views on policy options in each of the DSI topical areas. The goal in developing these papers was to identify and classify issues that effected each of the NRC programs and, ultimately, the means by which the agency gets its work done. The 16 DSIs were assembled in the *Strategic Planning Framework* (NRC, 1996), which was made available for public comment on September 13, 1996.

"DSI 5" applied to NRC's LLW program. The position paper superceded the staff's earlier 1993 program analysis by recommending six options for managing NRC's LLW programs. The six options proposed are as follows:

Option 1:   *The NRC assumes a greater leadership role in the National LLW program.*

Option 2:   *The NRC assumes a stronger regulatory role in the in the National LLW program.*

Option 3:   *The NRC retains the current LLW program.*

Option 4:   *The NRC recognizes progress in the National LLW program and reduces the size of its current program.*

Option 5:   *The NRC recommends to Congress that its LLW responsibilities be transferred to the EPA.*

Option 6:   *The NRC encourages the long-term storage of LLW under the concept of "assured storage."*

In a Staff Requirements Memorandum (SRM) dated March 7, 1997, the NRC Executive Director of Operations informed the staff of the Commission's preference for Option 3, to maintain the current LLW program. The ACNW provided it's views regarding DSI 5 and other cross-cutting issues outlined in the *Strategic Planning Framework* in a letter dated January 30, 1997. See Section 8.3.2 of this paper.

PART III: PAST ACNW ADVICE AND RECOMMENDATIONS

9 PREVIOUS ACNW REVIEWS

The ACNW was not in existence at the time NRC's LLW regulatory framework at Part 61 was created. Nevertheless, the Committee has commented on the implementation of that framework in more than 20 letter reports. The purpose of this section is to summarize past ACNW advice in the LLW area.

9.1 Background

The ACNW was established by the NRC in June 1988 as a Federal Advisory Committee to provide independent technical advice on agency activities, programs, and key technical issues associated with regulation, management, and safe disposal of certain types of radioactive waste. The Committee is independent of the NRC staff and reports directly to the Commission, which appoints its members. Consistent with NRC's regulatory mission, the ACNW undertakes independent studies and reviews related to the transportation, storage, and disposal of HLW and LLW, including the interim storage of SNF; materials safety; and facility decommissioning. The ACNW also independently evaluates staff efforts to develop and apply a risk-informed and performance-based regulatory framework to these programs (see Appendix B), consistent with Commission direction. This would include reviews of and comments on proposed rules, regulatory guidance, licensing documents, staff positions, and other issues, as requested by the Commission.

The operational practices of the ACNW are governed by the provisions of the Federal Advisory Committee Act (FACA – Public Law 92-463). FACA requires that, with very few exceptions, advisory committee meetings will be open the public.[59] The results of the ACNW's reviews, consisting of both comments and recommendations, are documented in letter reports. For the period 1988-2005, the ACNW has issued about 200 letter reports. Each year, ACNW letter reports are complied and published as updates to NUREG-1423 (ACNW, 1990–2005).

9.2 Discussion

Since its establishment, the ACNW has closely followed public health and safety issues associated with the management of LLW. Past ACNW letters may be generally classified as having been written in response to requests from the Commission, the Executive Director for Operations or NRC Program Office staffs although others have been prepared in response to a

---

[59] FACA requires that Committee memberships be fairly balanced in terms of the points of view represented and the agency functions being performed. As a result, members of specific advisory Committees tend to possess skills that parallel the program responsibilities of their sponsoring agencies.

perceived need which may have been identified by the staff, the public, licensees, or other agencies. In addition, the Committee had also closely followed international LLW practices and developments, as well as considerations arising from proposed or actual activities by the Agreement States. The Committee has held individual sessions, as well as Working Group meetings, dedicated to the actual licensing activities at proposed, as well as operating, LLW sites and their associated hearings.

Coupled with the broad experience represented by its membership and supporting staff, the ACNW's letters have covered a wide band of selected issues – groundwater monitoring, mixed LLW, onsite storage, performance assessment, and site characterization – in addition to specific technical topics such as the LLW source term and the suitability of certain types of LLW disposal containers. Also included in the Committee's deliberations have been broad topics concerned with the regulation of LLW and the associated NRC programs.

A list of past ACNW letters in the area of LLW can be found in Table 17. The Committee's first letter related to a LLW issue was written in August 1988. A brief summary of past recommendations from those letters is presented below.

9.3   Summary of ACNW Observations/Conclusions

The principal observations presented in Committee letters can be generally classified into six areas:

- General LLW Issues.

- Groundwater Monitoring Issues.

- Mixed LLW Issues.

- Onsite Storage Issues.

- Performance Assessment Issues.

- Comments on Waste Packages and Waste Form Issues.

9.3.1   General LLW Issues
Below Regulatory Concern Policy Statement. As noted earlier, LLWPAA required that NRC establish standards for determining when radionuclides are present in waste streams in sufficiently low concentrations or quantities as to be BRC, and therefore not subject to NRC regulation. As noted earlier, the Commission published its proposed policy statement outlining its plans to establish certain new BRC rules and procedures in August 1986 (51 FR 30839).

Table 17. ACNW Letter Reports Related to LLW Management. Listed chronologically. Electronic copies of these letter reports can be examined by going to the ACNW web site at *http://www.internal.nrc.gov/ACRS/rrs1/Trans_Let/index_top/ACNW_letters/ghindex.html.*

Letter Report Title	Date
ACNW Comments on Proposed Branch Technical Position Concerning Environmental Monitoring for Low-Level Waste Disposal Facilities	August 9, 1988
ACNW Comments on Proposed Commission Policy Statement on Regulatory Control Exemptions for Practices Whose Public Health and Safety Impacts are Below Regulatory Concern	August 9, 1988
Proposed Policy Statement on Below Regulatory Concern	September 15, 1988
Suitability of High Density Polyethylene Hing Integrity Containers	September 16, 1988
Final Rulemaking on 10 CFR Part 61 Relative to the Disposal of Greater-than-Class C Low-Level Radioactive Waste	February 24, 1989
Management of Mixed Hazardous and Low-Level Waste (Mixed Wastes)	May 3, 1989
Reporting Incidents involving the Management and Disposal of Low-Level Radioactive Waste	July 5, 1989
Comments on Technical Position Paper on Environmental Monitoring of Low-Level Radioactive Waste Disposal Facilities	September 19, 1989
Low-Level Waste Performance Assessment Methodology	October 18, 1989
NRC Program on of Low-Level Radioactive Waste	January 30, 1990
Regulation of Mixed Wastes	February 28, 1991
Comments Regarding 10 CFR Part 61 Proposed Revisions Related to Groundwater Protection	June 27, 1991
NRC Capabilities in Computer Modeling and Performance Assessment of Low-Level Waste Disposal Facilities	December 2, 1991
Proposed Expedited Rulemaking: Procedures and Criteria for On-site Storage of Low-Level Radioactive Waste	April 30, 1992
Source Term and Other Low-Level Waste Considerations	March 31, 1993
Review of Low-Level Radioactive Waste Performance Assessment Program	June 3, 1994
Private Ownership of Low-Level Waste Sites	February 6, 1995
Regulatory Issues in Low-Level Radioactive Waste Disposal Performance Assessments	June 28, 1995
Lessons-Learned from the Ward Valley, California, Low-Level Waste Disposal Facility Siting Process	August 10, 1995
Comments on SECY-95-201 and the NRC Activities Regarding Low-Level Radioactive Waste	December 29, 1995
Elements of an Adequate NRC Low-Level Radioactive Waste Program	July 24, 1996
Comments on Selected Direction-Setting Issues Identified in NRC's Strategic Assessment of Regulatory Activities	January 30, 1997
Time of Compliance for Low-Level Radioactive Waste Disposal Facilities	February 11, 1997
NRC Staff Research on Generic Posit-Disposal Criticality at Low-Level Radioactive Waste	July 30, 1998

Letter Report Title	Date
Branch Technical Position on Performance Assessment Methodology for Low-Level Radioactive Waste Disposal Facilities	August 2, 2000

The ACNW provided two sets of comments to the Commission on the proposed policy in letters dated August 9, 1988, and September 15, 1988. See Table 17. In 1993, the Commission withdrew its proposed policy.

Final Rulemaking on 10 CFR 61 Relative to the Disposal of Greater-than-Class-C Low-Level Radioactive Waste *(February 24, 1989)*. In 1988, the Commission proposed amendments to Part 61 that would require the deep geologic disposal of GTCC wastes unless an alternative means of disposal elsewhere was approved by the Commission. At its 7th meeting, in 1989, the Committee was briefed on the final proposed rule. Discussion's at the ACNW's meeting centered around the public comments received on the Commission's proposed rule (NRC, 1988; 53 FR 17709) and the staff's review and disposition of those comments. Following public comment, Part 61 was amended at §61.55(b)(2)(iv) to permit the disposal of GTCC waste in a HLW geologic repository licensed under Part 60 or some other type of disposal facility design approved by the Commission (NRC, 1989b; 54 FR 22578). Subject to certain recommendations, the Committee agreed with the final rule, as proposed by the staff. (Also see Section 6.5.1 of this paper.)

Reporting Incidents Involving the Management and Disposal of Low-Level Radioactive Wastes *(July 5, 1989)*. Previously, it had been observed that certain LLW form types had performed poorly in disposal facilities (e.g., NAS, 1976). To address this issue, recommendations were considered to characterize the various LLW streams to allow for the identification and treatment (stabilization) of problematic waste form compositions. The issue was addressed partially by the development of a staff technical position on LLW forms (NRC, 1991b). However, the Committee also believed that there should be a system for reporting performance incidents involving problematic LLW forms and it should be developed in a timely manner. The Committee was concerned that the limitations in staff resources at the time should be promptly addressed in development of such a system as a delay would be highly undesirable.

NRC Program on Low-Level Waste *(January 30, 1990)*. Earlier sections of this paper noted that by the early 1990s, the Commission's Part 61 regulatory framework was in-place supported by a considerable amount of staff guidance on how to implement that framework. For its part, consistent with direction from LLWPAA, DOE and EPA had also undertaken the development of additional technical information germane to the management of commercial LLW.

At its 16th meeting, the ACNW was briefed on the status of current LLW activities. As a result of that briefing, the Committee produced a letter with several recommendations. They first recommended that more attention should be given to the generator side of the LLW program with a focus processes effecting the types of LLW in the waste stream. The intent was to identify potential efficiencies in waste stream generation as a means of improving the management of LLW. The Committee believed that a "systems" approach to the management and disposal of LLW was necessary and could yield considerable dividends. (For NRC's part, they observed the need for closer coordination between the cognizant program offices with the agency.) They Committee also recommend the need for greater integration of all pertinent technical information from all cognizant agencies. They believed that a "road map", providing comprehensive guidance to licensees should be provided. Such guidance would be referenced, annotated, and contain key regulatory guides, technical positions, NUREGs as well as other technical information developed by other cognizant agencies. The Committee also recommended the preparation of a report that includes insights of current operating experience at existing LLW facilities with a view on how to improve NRC's regulatory responsibilities. Lastly, the Committee recommended that the Commission increase its efforts to accelerate the process for developing new disposal facilities.

Private Ownership of Low-Level Waste Sites (February 6, 1995). In 1994, the Commission issued a ANPR (59 FR 39485) that indicated that it was considering to allow private ownership of LLW sites as an alternative to the current requirement at §61.59(a) that permits only Federal or State ownership. The Committee concluded that there are no fundamental reasons why private ownership of LLW disposal sites should be prohibited but found several related issues, in its view, that required deliberate and cautious action.

The first major issue identified by the ACNW concerned the need for assurance of the protection of the health and safety of the public and the environment. During then-recent Commission policy discussions on adequacy and compatibility, the topic of provisions for private ownership of waste disposal sites was not included. The Committee expressed the view that the NRC needed to include explicit statements for pertinent requirements under the heading of adequacy and compatibility if the Commission proceeds with generic approval of private ownership. The Committee believed that NRC should require effective and timely transfer of ownership to another responsible and capable entity, such as the State, when any changes in the private ownership provision for waste sites, including dissolution of the corporate entity, are effected. The measure of adequacy and compatibility of Agreement and State operations should include effective and frequent monitoring and evaluation of private entities that are responsible for waste sites.

The Committee noted that §61.7(a) of the regulation presents 500 years as the target reference for siting and intruder barrier considerations. However, disposed LLW may pose a significant hazard for periods that, under some conditions, may well exceed 500 years. The Committee expressed the view that the Commission should expand the criteria to ensure that the State [or some governmental entity] maintain an active interest in the protection function of the disposal site for as along as the waste poses a hazard in the regulatory sense.

The second major issue concerned the administrative procedures that lead to privatization. The openness procedures used by the NRC to conduct its regulatory affairs provide ample opportunity for all interested parties to have their views considered. The Committee observed that given the potential importance of transferring LLW management accountability to a private corporate entity, with a likely modest life expectancy, compared to the period of time the waste possess a hazard, requires administrative (licensing) procedures comparable to those already used by the Commission. The Committee noted thus far it had not obtained information that this was the case when the State of Utah first acted.[60]

In summary, although the Committee believed that private entities were potentially capable of meeting the long-term protection function requirements of LLW management, final accountability for the long-term performance of a LLW disposal facility should continue to be through some type of governmental oversight entity. Furthermore, the Committee believed that the privatization decision-making process should be an open process not unlike the current administrative decision-making process already used by the NRC.

Following review, the Commission decided to not to amend §61.59(a).

Comments on SECY-95-201 and the NRC Activities Regarding Low-Level Radioactive Waste *(December 29, 1995).* In a September 14, 1995, SRM, the Commission requested the ACNW to provide comments on SECY-95-201 (NRC, 1995c), including practicable alternatives to the proposed options and the ACNW views on the significant consequences on the alternatives available to the Commission. See Section 7.2 of this paper.

SECY-95-201 identified three options regarding the future of the NRC LLW program. Briefly described, these options were:

- Continue the program as currently in place – *Option 1.*

- Reduce the program by eliminating or reducing various parts – *Option 2.*

- Terminate all parts of the LLW program – *Option 3.*

SECY-95-201 concluded that, based on statutory requirements and budget restrictions, Option 2 was the only practicable alternative. The Committee was unable to evaluate in detail the program as outlined in Option 2 because of the lack of specificity in resource allocations for various activities. The ACNW had a number of concerns with the conclusions of SECY-95-201. While current budgetary constraints were recognized, the Committee concluded that it is in the

---

[60]Acting in its capacity as an NRC-approved Agreement State, the State of Utah issued an exemption to governmental land ownership requirement in its LLW regulations to Envirocare of Utah in March 1991 when the State issued a license to that private corporation to allow it to operate a LLW disposal facility on privately-owned land.

national interest to have a centralized LLW program within the NRC and it strongly recommended that the Commissioners prioritize the LLW program in relation to all activities within the agency. Further, the Committee noted that the use of terms such as "limited" and "essential" to describe the resources and activities under Option 2 in SECY 95-201 was considered ambiguous. The Committee felt that the most important shortcoming of the SECY paper was its failure to address the fundamental question of what type of LLW program would be necesssary and sufficient to satisfy NRC's public health and safety mission.

Later in 1995, SECY-95-201 was overtaken by the Commission's *Strategic Assessment and Rebaselining Imitative* described in Section 7.2 of this paper.

Elements of an Adequate NRC Low-Level Radioactive Waste Program *(July 24, 1996)*. The Committee prepared this letter report in response to a request from then-Chairman Jackson as to the Committee's review of what would constitute an adequate LLW program. This topic was previously discussed in connection with the Committee's earlier review of SECY-95-201.

In its July 1996 letter, the Committee expressed the view that an adequate NRC LLW program was one that would ensure that the processing, storage, and disposal of LLW, as defined in Part 61, would be carried out in accord with other NRC regulations (e.g., Part 20) and that the current and future impact of such activities would not represent an excessive risk to the affected population or the environment. Further, the Committee observed it would also be desirable to include in such a program attention to GTCC LLW waste as defined in Part 61 and to "mixed waste." Under such an expanded scope, other wastes that would be included are: naturally occurring and accelerator produced radioactive material (NARM) and NORM, wastes from uranium recovery and processing, wastes that are formed by the inadvertent concentration of contaminants (e.g., sewage, bag house dust), and wastes derived from decontamination and decommissioning activity.

Comments on Selected Direction-Setting Issues Identified in NRC's Strategic Assessment of Regulatory Activities *(January 30, 1997)*. As noted earlier in this paper, the Commission under took a four-phase strategic planning exercise in 1995 known as the *Strategic Assessment and Rebaselining Imitative*. The principal focus of the initiative were the identification of key strategic issues associated with NRC's primary responsibility to protecting public health and safety and the environment. These key issues were called DSIs and DSI 5 applied to NRC's LLW program. The ACNW provided it views regarding DSI 5 and other cross-cutting waste management issues outlined in the *Strategic Planning Document* in a letter dated January 30, 1997 (Pomeroy, 1997).

In its letter, the ACNW recommended that the Commission adopt Option 2 – "Assume a strong regulatory role in national LLW program." The Committee had other recommendations, including:

- A number of waste types were missing from the discussion. In its general introductory comments, the Committee noted its concern about the omission of DSI cross-cutting issues such as the management of mixed wastes and GTCC LLW. The Committee believed these issues need to be adequately addressed in the strategic planning of the agency.

- NRC's acceptance of long-term storage of LLW, although attractive as a practical solution to a current problem, may not be acceptable to the Nation. The current national policy is to provide final disposal by the present generation in a manner that does not jeopardize public health and safety now and in the future. The DSI paper did not adequately address the requirements for implementing long-term storage of commercial LLW. The Committee was also concerned about the rather favorable light placed on interim storage in the DSI paper presumably because to date no incident has been reported as a result of storage on the originating site. However, no evidence exists that onsite storage can be effective over the expected life of the waste and the proliferation of storage sites enhances the risk.

- The Committee suggested that caution be exercised in using "rules of thumb" to define waste types in terms of the length of time over which they may be hazardous. In view of the absence of a *de minimis* position regarding radioactivity and the broad application of the no threshold-linear view of the health effects of radiation, the Committee suggested rules of thumb are a significant oversimplification.

- Finally, the Committee questioned the acceptance of DOE waste sites as potential disposal sites for civilian wastes. Existing DOE sites were not selected on the basis of criteria used in siting and licensing civilian disposal facilities, and evidence is lacking that these sites could meet the standards and regulations in effect.

In conclusion, the Committee recommend Option 2 of this DSI paper but encourage additions to (a) develop a more comprehensive definition of LLW and (b) evaluate the potential implementation and impact of assured storage with adequate protection and termination procedures.

9.3.2   Groundwater Monitoring Issues
ACNW Comments on Proposed Branch Technical Position Concerning Environmental Monitoring for Low-Level Waste Disposal Facilities (August 5, 1988). In November 1987 (52 FR 42486), the NRC staff made a Branch Technical Position (BTP) available for public

comment. The BTP addressed the Part 61 Subpart D requirements at §61.53(c) for environmental monitoring of LLW disposal facilities. Following the request for comments, work on the BTP was interrupted because of resource limitations. In its comments to the Commission, the ACNW recommended that work on the BTP should be completed and that the guidance be issued in final form. However, in making its recommendation, the Committee also recommended that the overall purpose of the staff's technical position in this area needs to be clarified, specifically to indicate whether it is prepared to provide guidance on monitoring policy or to prescribe detailed monitoring requirements.

Comments on Technical Position Paper on Environmental Monitoring of Low-Level Radioactive Waste Disposal Facilities *(September 19, 1989)*. After a brief interruption, staff work on the development of a BTP on environmental monitoring of LLW disposal facilities continued, but under the new title of a *Technical Position Paper*. Following a second review, the Committee believed that the renamed paper was acceptable for publication. However, the guidance document was never finalized but was later identified as a candidate area by the staff for rulemaking. See Table 16. As noted earlier (Section 8.2 of this paper), in recent years, RES has sponsored many research projects and public workshops related to the subject of environmental monitoring and modeling.

Comments Regarding 10 CFR Part 61 Proposed Revisions Related to Groundwater Protection *(June 27, 1991)*. In a September 6, 1990, letter, the ACNW recommended that the revised NRC technical position on waste form (NRC, 1991b) be published in final form. Along with the recommendation, though, the Committee expressed several concerns, including the need to revise Part 61 to show more direct emphasis on the resistance of LLW forms to leaching by percolating groundwater. In a December 31, 1990, SRM, the Commission requested that the Committee justify its position by evaluating the efficacy of the existing Part 61 in meeting its concerns.

In a subsequent meeting with staff, the history and performance experiences of earlier NRC-licensed LLW disposal facilities (Table 2) was reviewed, particularly as it related to the migration of radioactive materials. It was noted that the staff considered this past experience in scoping the Part 61 EIS and developing rule subsequent LLW regulation. The Committee also were apprised of the staff efforts at the time to undertake detailed studies of contaminant flow and transport phenomena as part of a broader LLW performance assessment effort (later to be documented in NUREG-1573 (NRC,2000). (See Appendix D of this paper for a more detailed discussion of this NUREG.) Based on this emerging work, the Committee was assured that it would provide additional insights into groundwater protection issues. Lastly, the Committee held a "brainstorming session" with NRC staff and their technical assistance contractors at the time which explored options that might improve radionuclide retention in, or to retard radionuclide migration from LLW forms.

On the basis of these interactions, the Committee set aside its suggestion that Part 61 be revised to explicitly include a requirement for LLW waste performance as a means of enhancing ground water protection.

9.3.3   Mixed LLW Issues

Management of Mixed Hazardous and Low-Level Radioactive Wastes (Mixed Wastes) *(May 3, 1989)*. Although not addressed in this paper, chemically-hazardous LLW is subject to dual regulation under EPA's RCRA regulation. Following meetings with the NRC staff and representatives from the Nuclear Management and Resources Council (NUMARC), the ACNW offered several recommendations to the Commission. The Committee believed that additional resources should be assigned to study this issue, that its resolution was primarily institutional, and that the problems caused by dual jurisdiction are solvable (although at the time it seemed to be recognized by most knowledgeable institutions that any facility meeting NRC regulatory requirements is capable of meeting EPA criteria for the disposal of hazardous [nonradioactive wastes]).

The Committee also observed that the management of chemically hazardous GTCC LLW, NARM, and NORM is an area that had been overlooked and recommended attention by the staff.

Regulation of Mixed Wastes *(February 28, 1991)*. Following the ACNW's May 1989 letter, OTA (1989) published a comprehensive report on the status of the national LLW program. That report also included an examination of mixed LLW issues and in doing so noted that the lack of mixed waste treatment options, access to mixed waste disposal facilities, and conflicting (and inconsistent) EPA and NRC regulations. At the request of then-Commissioner Curtis, the ACNW reviewed the comparability of protection afforded by NRC and EPA regulations when applied to the disposal of mixed wastes. The Committee responded to the request by conducting a Working Group Meeting devoted to the subject in December 1990 as well as dedicating addition time to the matter at subsequent Committee meetings.

Following on to the previous may 1989 ACNW letter on mixed LLW, the Committee reported that an industry-sponsored study (NUMARC, 1990) seemed to indicate that a facility built in conformance with Part 61 was slightly superior to a facility built in conformance with EPA's RCRA regulations at 40 CFR Part 264. However, the NRC staff stated that certain features of the disposal facility designed to those regulations, such as the requirement for a double liner and the leachate collection and retention provisions, "...appear to offer enhanced protection of groundwater, at least temporarily...." The Committee also noted that the then proposed EPA LLW standard (Part 193) included a "... subsystem requirement that groundwater contamination be limited so that no offsite person will receive an effective dose rate greater than 0.04 mSv (4 mrem) per year, may be a potential important attribute of the EPA regulations that is important...." Several other considerations were discussed in the ACNW's 1991 letter. It was

NRC FORM 335 (9-2004) NRCMD 3.7	U.S. NUCLEAR REGULATORY COMMISSION	1. REPORT NUMBER (Assigned by NRC, Add Vol., Supp., Rev., and Addendum Numbers, if any.)
	BIBLIOGRAPHIC DATA SHEET *(See instructions on the reverse)*	NUREG-1423, Volume 15

2. TITLE AND SUBTITLE	3. DATE REPORT PUBLISHED	
A Compilation of Reports of the Advisory Committee on Nuclear Waste - September 2004 - June 2006	MONTH	YEAR
	October	2006
	4. FIN OR GRANT NUMBER	

5. AUTHOR(S)	6. TYPE OF REPORT
	Compilation
	7. PERIOD COVERED *(Inclusive Dates)*
	September 2004 - June 2006

8. PERFORMING ORGANIZATION - NAME AND ADDRESS *(If NRC, provide Division, Office or Region, U.S. Nuclear Regulatory Commission, and mailing address; if contractor, provide name and mailing address.)*

Advisory Committee on Nuclear Waste
U. S. Nuclear Regulatory Commission
Washington, DC 20555-0001

9. SPONSORING ORGANIZATION - NAME AND ADDRESS *(If NRC, type "Same as above"; if contractor, provide NRC Division, Office or Region, U.S. Nuclear Regulatory Commission, and mailing address.)*

Same as above

10. SUPPLEMENTARY NOTES

11. ABSTRACT *(200 words or less)*

This compilation contains 30 reports issued by the Advisory Committee on Nuclear Waste (ACNW) during the sixteenth and seventeenth years of its operation. The reports were submitted to the Chairman and the Executive Director for Operations of the U. S. Nuclear Regulatory Commission (NRC). Reports prepared by the Committee have been made available to the public through the NRC Public Document Room, the U. S. Library of Congress, and the Internet at http://www.nrc.gov/reading-rm/doc-collections.

12. KEY WORDS/DESCRIPTORS *(List words or phrases that will assist researchers in locating the report.)*	13. AVAILABILITY STATEMENT
	Unlimited
Nuclear Waste Management High-Level Radioactive Waste Future Volcanism Igneous Activity Low-Level Radioactive Waste Safety Engineering Safety Research Yucca Mountain	14. SECURITY CLASSIFICATION
	(This Page)
	Unclassified
	(This Report)
	Unclassified
	15. NUMBER OF PAGES
	16. PRICE

NRC FORM 335 (9-2004)

PRINTED ON RECYCLED PAPER